Rik Das, Siddhartha Bhattacharyya, Sudarshan Nandy (Eds.)
Machine Learning Applications – Emerging Trends
De Gruyter Frontiers in Computational Intelligence

De Gruyter Frontiers in Computational Intelligence

Edited by Siddhartha Bhattacharyya

—

Volume 5

Machine Learning Applications

Emerging Trends

Edited by
Rik Das, Siddhartha Bhattacharyya, Sudarshan Nandy

DE GRUYTER

Editors

Rik Das
Xavier Institute of social services
Dept. of Information Technology
Dr. Camil Bulcke Path (Purulia Road)
RANCHI-834001 , India
rikdas78@gmail.com

Prof. (Dr.) Siddhartha Bhattacharyya
CHRIST (Deemed to be University),
Dept. of Computer Science and Engineering
Bangalore-560029, India
dr.siddhartha.bhattacharyya@gmail.com

Dr. Sudarshan Nandy
Amity University
Amity School of Engineering and Technology
KOLKATA-700135, India
sudarshannandy@gmail.com

ISBN 978-3-11-077705-5
e-ISBN (PDF) 978-3-11-061098-7
e-ISBN (EPUB) 978-3-11-060866-3
ISSN 2512-8868

Library of Congress Control Number: 2020935231

Bibliographic information published by the Deutsche Nationalbibliothek
The Deutsche Nationalbibliothek lists this publication in the Deutsche Nationalbibliografie; detailed
bibliographic data are available on the Internet at http://dnb.dnb.de.

© 2021 Walter de Gruyter GmbH, Berlin/Boston
This volume is text- and page-identical with the hardback published in 2020.
Printing and binding: CPI books GmbH, Leck
Cover image: shulz/E+/getty images

www.degruyter.com

Dedication

Rik Das would like to dedicate this book to his father Mr. Kamal Kumar Das, his mother Mrs. Malabika Das, his better half Mrs. Simi Das and his kids Sohan and Dikshan

Siddhartha Bhattacharyya would like to dedicate this book to his father Late Ajit Kumar Bhattacharyya, his mother Late Hashi Bhattacharyya, his beloved wife Rashni and his cousin sisters-in-law Nivedita, Madhuparna, Anushree and Swarnali

Sudarshan Nandy would like to dedicate this book to his father Mr. Sukumar Nandy, his mother Mrs. Anjana Nandy, his loving wife Mrs. Surabhi Nandi and his son Samarjit

https://doi.org/10.1515/9783110610987-001

Preface

Machine learning is a way to infuse intelligence in data analytics. With the growth of civilization, researchers have invested efforts to mimic human traits in designing intelligent algorithms for solving real world problems. This volume is intended to give a bird's eye view of the latest trends in computationally intelligent algorithms and devices helpful for the mankind. This volume would benefit engineering students, tutors, research scholars and entrepreneurs with enriched interdisciplinary approaches to find new ways of applications in machine learning.

The volume is attempted to address emerging trends in machine learning applications. Recent trends in information identification have identified huge scope to in applying machine learning techniques for gaining meaningful insights. Random growth of unstructured data poses new research challenges to handle this huge source of information. Efficient designing of machine learning techniques is the need of the hour. Recent literature in machine learning has emphasized on single technique of information identification. Huge scope exists in developing hybrid machine learning models with reduced computational complexity for enhanced accuracy of information identification. This book will focus on techniques to reduce feature dimension for designing light weight techniques for real time identification and decision fusion. Key Findings of the book will be the use of machine learning in daily lives and the applications of it to improve livelihood. However, it will not be able to cover the entire domain in machine learning in its limited scope. This book is going to benefit the research scholars, entrepreneurs and interdisciplinary approaches to find new ways of applications in machine learning and thus will have novel research contributions. The lightweight techniques can be well used in real time which will add value to practice.

The volume contains seven well versed chapters entailing recent machine learning techniques and applications.

Humans are considered as the smartest animal because of their capacity to solve complex jobs. These various tasks can be divided into two different classes, namely, simple tasks and complicated tasks. The categorization depends on the complexity of the task level. The complexity of a task can be measured and it depends on the steps through which the task is solved. It can, therefore, be said that some suitable sequential steps are necessary to finish a specific assignment. However, appropriate data is required to achieve this objective. At the same moment, human attempts to learn and obtain other associated data or features that can effectively help enhance a job. Hence, suitable information along with previous experiences can help to complete a task more efficiently with less number of errors or mistakes. With this analogy to human beings, the capability of a system can be improved through learning from the continuous process of observation and past experiences which is itself a repetitive task. Chapter 1 provides a brief introduction to the subject area along with highlights of recent research trends.

https://doi.org/10.1515/9783110610987-002

Rapid use of social media for communication and information sharing, use of world wide web as a huge information repository, advancement in data and image capturing techniques have drastically increased the volume and size of the data. Analyzing, processing and searching such a huge volume of data is a complex task. This also reduces the accuracy as well as diminishes performance. To overcome these problems dimension reduction techniques are used to reduce the size of data and features without the loss of valuable information. Dimension of features can be reduced by projecting the features to a low dimensional space or using feature selection or feature encoding techniques. These techniques can be linear or non-linear depending on the relationship between data in higher and lower dimensions. In Chapter 2, a detailed analysis has been performed on some of the non-linear dimensionality reduction techniques and their applications.

With the growth of the internet and social media, music data is growing at an enormous rate. Music analytics has a wide canvas covering all aspects related to music. Chapter 3 provides a glimpse of this large canvas with sample applications covered in detail. Machine learning has taken a central role in the progress of many domains including music analytics. This chapter will help the readers to understand various applications of machine learning in computational musicology. Music feature learning and musical pattern recognition give conceptual understanding and the challenges involved. Feature engineering algorithms for pitch detection or tempo estimation are covered in more detail with available popular feature extraction tools. Music classification and clustering examples explore the use of machine learning. Various applications ranging from the query by humming to music recommendation are provided for efficient music information retrieval. Future directions and challenges with deep learning as a new approach and incorporation of human cognition and perception as a challenge make this domain a challenging research domain.

Character recognition is a challenging area in Machine Learning, Pattern Recognition or Image Processing. The accuracy to recognize handwritten character by human is far better compared to machine recognition. To develop an interface which can differentiate characters written by human yet requires intensive research. Though number of researches have presented in this area, still research is going on to achieve human like accuracy. Both handwritten and printed character recognition are categorized into two types, online and offline. A good number of researches have done work in the area of optical character recognition in different languages but for the Odia language, development is negligible. Odia (formerly it was Oriya), one of the 22 scheduled language recognized by the constitution of India and it is the official language of the state of Odisha (Orissa), more than 40 million people speak Odia. Due to the roundish shape of Odia character, large number of modified and compound characters and similarity between different characters makes this language very hard to create a satisfactory classifier. In the present survey undertaken we have discussed what are challenges be for Odia language and the machine learning techniques used in the recognition of Odia character recognition. Chapter 4 describes the complete process of character recognition i.e. pre-processing, extraction and selection of feature set and character recognition elaborately with comparison analysis and the metrics used to evaluate machine learning algorithms.

Recommendation systems have been the integral part of web and mobile applications in the domains of e commerce, e learning, e health, e governance, social networking and search engines. The problem of spending more time in getting the required relevant information from the many options available, also called as the information overload problem is addressed by the recommendation systems. The context based recommendation systems are the types of recommendation systems which use the context information to provide the recommended items. The context is the data about the application or the surroundings or the purpose with which the user is interacting with the system which can be like time, location, type of product, user's purpose or any situation describing the interaction. In Chapter 5, the architecture of the context based recommendation system is proposed with the pre filtering method with context rules. The class and object based model of context with rules and recommendation system is proposed which can be converted into a relational model for data storage for the recommendation system. One of the actions like rating is used to predict the preference of items for the current user. The analysis of the system is carried out with the Het Rec 2011 Movielens data set. The accuracy measure MAE (Mean Absolute Error) is analysed in the proposed work and the relevance is measured in terms of precision, recall and F1 measure. The experimental result shows the influence of context on action and improvement in quality of recommendation with the proposed method.

In some underdeveloped and developing country, crop disease detection depends on the field experience of the farmer and this may cause degradation in production quality. The production quality can be improved, if the diseases detected in an earlier stage and machine-learning techniques are applied for detection of the crop diseases. The implementation of the decision support system is also an important part because the farmer needs to know proper information in a real-time manner and hence proper action can be taken in an earlier stage. Several advanced techniques are proposed for early detection of the crop diseases and it is inspired to survey on the techniques of machine learning based decision support system, address the issues related to productivity in agriculture. Chapter 6 analyzes and discusses the specific model of machine learning techniques. Our finding indicates the justification and efficiency of incorporating the machine learning techniques with a decision support system for early detection of crop diseases.

Chapter 7 concludes with a discussion on the takeaways from the volume. This chapter also focuses on the future directions of research in the machine learning domain.

This volume has readily addressed all these shortcomings of the existing competing titles and has proposed chapters to identify and solve real time problems raised due to lack of feature dimension reduction. An extra characteristic has been added to this volume with introduction of hybrid machine learning. Reduced dimension of multiview features will encourage various fusion techniques and will enhance the accuracy of identification. Thus the volume qualifies to be a work related to current state of problem definition and will definitely show new paradigms for information identification using machine learning techniques.

Kolkata, India
September, 2019

Rik Das
Siddhartha Bhattacharyya
Sudarshan Nandy

Inhalt

Inhalt

Siddhartha Bhattacharyya, Rik Das, and Sudarshan Nandy

1 A prologue to Emerging trends in Machine Learning

1.1 Introduction

Humans are considered as the smartest animal because of their capacity to solve the complex job. These various tasks can be divided into two different classes, namely, simple tasks and complicated tasks. The categorization depends on the complexity of the task level. The complexity of a task can be measured and it depends on the steps through which the task is solved. It can, therefore, be said that some suitable sequential steps are necessary to finish a specific assignment. However, appropriate data is required to achieve this objective. At the same moment, human attempts to learn and obtain other associated data or features that can effectively help enhance a job. Hence, suitable information along with previous experiences can help to complete a task more efficiently with less number of errors or mistakes. With this analogy to human beings, the capability of a system can be improved through learning from the continuous process of observation and past experiences which is a repetitive task [1] [2].

The learning method is split into three distinct classes based on the learning strategies. These include monitored learning, unattended learning and enhancement learning. The first way of learning is with the guidance of experts. This sort of learning has a primary motive to receive advice from specialists directly. In monitored education or supervised learning, the system predicts the group of similar type unknown entities based on past knowledge. Supervised learning is guided by the intelligence which is achieved from the supervisors or experts. In the indirect learning process, individual perception develops indirectly. In unsupervised learning, system or environment examines the suitable patterns of unknown entities by clustering similar entities together. The third type of learning is self-learning. In this learning process, system or environment learn from its self-knowledge or self-experience. In the reinforcement learning process, the learner learns based on the knowledge gained from several efforts or attempts even if some of the attempts fail [3] [4] [5].

Machine learning process attempts to replicate the procedure of humans learning. Moving to the machine-learning procedure, the huge pool of information is available from input data. Due to the abundance of data in current circumstances, machines should learn the handling of data automatically and discover unrecognizable patterns without human intervention. Machine learning is defined as the ability of a system to predict a pattern in available data without being explicitly programmed [5] [6] [7]. Human beings have limited efficiency in handling a huge amount of data due to an intrinsic manual error in analysing the same. As a result, machine learning and deep

https://doi.org/10.1515/9783110610987-003

learning has gained immense popularity as an active area of research in recent times. The scenario is generated due to the easy availability of data from assorted sources including the Internet, social media, public datasets, and gadgets and so on. All these available data require huge databases to be stored, archived, maintained and identified for potential usage. Most of the online and offline business houses have adopted data-driven promotion of their products and services. Thus, data is the lifeline of contemporary existence and has a pivotal role in revenue generation in addition to lead identification. Machine learning environment utilizes a set of characteristics based on common characteristics that assist define each individual object within a dataset. Failure to select characteristics can lead to errors during the evaluation phase and reduce the learning environment's effectiveness [8] [9].

1.2 Literature Survey

1.2.1 Supervised learning

Let us consider a learning environment in which the machine acquires countless fruit pictures as an input. The primary job of the machine is to separate the fruits by category. The pictures can be segregated by two primary characteristics such as form and colour. If the pictures are separated by form, it is necessary to separate a round-shaped fruit from a square-shaped fruit. If segmentation requires a colour-based classification, then a yellow-coloured image is separated from a green coloured image. The issue occurs, however, how a non-human system distinguishes the shape and colour of the pictures. At the very beginning, the stage of a learning environment of a machine is similar tothe child teaching-learning process. A child does not identify any object without guidance. Likewise, without basic suitable knowledge or expert advice, a learning environment cannot start to work. A system requires some basic information that will be provided to it. In machine learning methodology, the basic information or expert information of an object is extracted from the training dataset. Training data is relevant information about a particular object or task. These data specifies different tags, which have multiple past information on numerous aspects and features. This information is collected from n numbers of input images. Sometimes, tags are also known as labels. So, training data are labelled in supervised learning [10] [11] [12] [13].

The supervised learning process involves many significant steps. At the initial step, a dataset is required for the task. Datasets need to be well organized and a complete one. A well-known dataset is generated based on the suggestions of experts. An expert can suggest which information (properties or features) are most suitable for the task. On the other hand, if an expert is not available, then „brute-force" is the easiest way to do the task. However, in this approach, datasets are not ready for directly use [14] [15]. It is being observed that datasets may contain many noises and missing data. For that reason, systems require a step, which should pre-process data for finding noisy and missing data. Hence, the next step is data construction and data pre-processing. Several methodologies exist for data pre-processing to handle noisy and missing data. In the case of the large data set, along with the problem of handling noise and missing data, data optimization problems are also important. Optimization algorithms minimize data and maintain the quality in the large dataset for working effectively. Various algorithms exist for sampling and minimizing the data from a large dataset for different optimization problems [16] [17] [18]. If once your dataset is pre-processed, then the next task is feature selection. It is a procedure of finding and elimination of irrelevant and duplicate features as many as possible. Many informative features selected and constructed through data mining algorithms together increases the efficiency and accuracy of the supervised learning process. Creating new features from an old or basic feature set is known as feature construction or transformation. These freshly constructed meaningful features are very valuable for better understanding and producing better accuracy with unambiguity for classifier [19] [20]. Selection of algorithms is a significant step in supervised learning. The classifier assessment is more acknowledged based on the precision obtained for a model of the forecast. Mostly, it uses three kinds of methods to identify a classifier's precision. Dividing the instruction set into two components is one method i. e. two-third patterns is allowed for the training set and rest of the patterns are used for the test set. Cross-validation is another method where the dataset is to be divided into two subsets of the same size in a mutually exclusive way in this operation. Based on the mixture of all other subsets, each subset is taught by the classifier. This classifier calculates its error rate by estimating the average error rate that is measured from the error rate of the individual subset. In respect to computational time, this methodology is expensive but it is more popular when the correct classifier error rate is required. If the error rate is too high, then the classifier must return to the monitored machine learning method to some or some previous step. To discover the cause of elevated error rate, various variables need to be examined such as not using suitable characteristics for the job, training set size is not enough, problem dimensionality is too big, algorithm choice is incorrect or some modification of parameters is needed. Standard and most commonly used supervised learning algorithms are K-nearest neighbour (k-NN), Decision tree, Linear regression, Logistic regression, Support Vector Machine (SVM), etc. [21] [22] [23] [24] [25].

Several supervised machine-learning methods, such as logical or symbolic methods, perceptron-based methods and statistical techniques, are introduced in artificial intelligence systems. [27]. Examples of the two most significant symbolic or logical methods are decision trees and classifiers based on rules. Feature values play a major part in grouping and abbreviating items in the decision-making trees. When a learning scheme is interested in using multi-dimensional multi-class assessment, the decision tree is the best option. This learning method aims primarily to develop a new approach that predicts input values. In the decision tree, each node reflects a vector of a function and each edge displays a feature value. Depending on the values of the function, objects start grouping and arranging from the root node in sorted order. The main issue of developing optimal binary trees is an NP-complete problem. Therefore, researchers are finding efficient heuristics for developing near-optimal decision trees. Training information is split and defined based on the function in the root node of a decision tree. The methods that split the training data are information gain and Gini index. A choice tree, or any learned hypothesis k, is regarded as over-fit training information when a distinct assumption k′ happens that has more mistake than h when checked on the training information, but a lower error than h when checked on the full information. To avoid overfitting training data, decision tree classification algorithms follow two popular methods: I first, if a training algorithm fails to correctly achieve the likely fit point, then the training algorithm is stopped. (ii) Second, the decision tree is pruned. Now the pruning is performed by considering the tree with larger leaves because the tree with the larger leaves usually provides optimum accuracy. Pre-prune is the best upfront method for managing to fit. The decision tree, however, does not allow it to grow to its full dimension. The threshold evaluation for the metric function value can result in a non-trivial goal being terminated. Decision tree classifiers usually participate in post-pruning activities that evaluate the efficiency of decision trees as they are thinned through the use of a validation set. The extreme prevalent class of the training dataset that is structured for it can be extracted and assigned to any node. There is a sensible overhaul of known pruning methods that there is no separate best pruning method. Decision trees can be converted into a collection of rules by producing a separate rule from the root to a leaf in the tree for individual routes. However, instructions can also be drawn from training data using a rule-based algorithm heterogeneity [28] [29] [30] [31] [32] [33].

The goal of supervised learning is to create the minimum number of rules that are unswerving with the training information. A large number of well-known rules generally indicate that the learning algorithm is intended to „remember" the collection of instruction instead of defining the laws that direct it. A rule is inspected by a divide-and-conquer algorithm that describes a part of its training instances. Now split these cases and recurrently overcome the remaining by learning additional rules until there are no continuing instances. There is a universal pseudo-code for apprentices of the rule. The difference between heuristic of understanding rule and heuristic of the decision tree is that finishing calculates the average value of a collection of disorganized sets. Each individual function value is checked while the apprentices of rule evaluate only the quality of the set of instances included in the applicant rule. Further forward-looking rule apprentices diverge from this simple pseudo-code generally by totalling supplementary mechanisms to avoid over-fitting of the training data. The specialization method ends with quality measurement or in a discrete pruning period by simplifying more than usual rules for competence. Consequently, creating rules of judgment with exceptional expectations or consistency is important for a rule learning scheme. Generally speaking, these characteristics are evaluated using a function called rule quality. The amount of rule quality is sought— for both the procedures of rule practice and classification. A rule quality is calculated in rule training and can be used in the rule description and/or simplification process as a norm. In classification, the quality value of a rule can be associated with each rule to determine discrepancies when the cases to be categorized comply with several regulations [34] [35] [36].

1.2.2 Perceptron-based techniques

Perceptron can categorize object sets that are linearly separated. The input objects are linearly separable if it is possible to draw a straight line between two different classes. Now if it is a case where the classes of the given dataset are not separable then training with any learning algorithm for a perceptron is never fully classified a dataset. There are two kinds of perceptron-based methods: single-layer perceptron and multi-layer perceptron [27] [37].

A perception of a single layer is very easy. The input layer and output layer comprises of two levels. Take into account that the values in the input layer are x1, x2, x3,...., xn and link masses are w1, w2,..., wn. Weights are normally true numbers in the interval[-1, 1]) whereby the perceptron calculates the sum of weighted inputs. If the summation exceeds the limit, the production is zero. The best collective strategy is to learn from the compilation of training information via the perceptron operation. This method repeatedly executes the algorithm via the training set up, to discover a precise preview vector for the full training set [38] [39].

A neural multi-layer network has enormous neurons coupled in a network structure. In a network, neurons are usually split into 3-type input neurons, which allow processing information; output neurons, in which processing results occur; and concealed neurons in the a3-type neuron. In the neural network, the flow of processing usually happens from the input layer to the output layer and weights in between the layers are updated accordingly. Ultimately the modification of well-connected weight of each layer in the neural network helps to recognise the dataset. The indication on the input neurons will broadcast the whole mode through a network during cataloguing or grouping so that the initiation values are regulated for each neuron input. The initiation value for input neuron indicates almost an internal function of the network. Each input neuron then sends its initiating value to each invisible or concealed neuron that is linked. These concealed neurons, respectively, calculate their individual initiation value and then die to the resulting neurons. The initiation value is evaluated according to a modest initial feature for every receiving neuron. The functional calculations gather the support of all neurons when the participation of neuron is demarcated by the significance of initiation of the neurons that are sent. This addition is then usually modified by a subsidiary, e. g. by controlling the value-added between zero and one. Normally there is a problem with correctly defining the number of hidden layers because a wrong calculation of the size of the neurons can lead to inaccurate estimations and simplification, whereas unnecessary nodes can overfit and eventually create a further general optimum challenge. There are three key features which depend on ANN. These are the starting features of the input neuron, the network design and the weight of the input link. The original two features have to be fixed; the ANN's activities are demarcated by the current weight values. The network weights are mainly allocated to random values, which are then repeatedly displayed in the training set. The instances of a pattern are set to the input layers of a neural network and the error is calculated by comparing the actual outcome and expected output. Then, each network weight is modified in a manner that conveys the network output values closer to the desired output values [39] [40] [41] [42] [43] [44].

In literature, there are different training algorithms available and back-propagation (BP) is one of the famous training algorithms for the neural network. There are six successive phases in the BP algorithm. In the first phase, the ANN (Artificial Neural Network) is able to arrange sampling information. The second stage consists of comparing the net output with the desired test value and measuring the output neuron mistake separately. In the following stage, calculate the importance of the production and scaling variable for each neuron to match the limit value to the output. It causes a local mistake. Change the neuron weight individually to address the local mistake. In the following phase, the local fault is assigned to the detection of the neurons at the prior stage. It makes the cells connected by greater weights more responsible. Ultimately, repeat the overhead stages of the cells at the above stage, with individual „fault" as its error. Before it reaches a straight weight system, the backpropagation algorithm shall adjust the weight. Each repeat/epoch of the learning phase requires $O(nW)$ time for n teaching cases and W weights. Depending on the number of outputs, the number of epochs can be improved exponentially. Neural networks, therefore, use a distinct amount of ending laws when learning ends. There are some final laws that collectively lead. Firstly, after a quantified amount of epochs, the teaching method will be terminated. Secondly, the coaching method is stopped if an error is calculated and a limit is achieved. The learning method is subsequently stopped if there is no evolution of error calculation throughout a certain period. Finally, if the error calculated from the training data on studied information is equivalent to the positive amount, as opposed to the error calculated on the training set, the coaching technique is halted. The innovative algorithm for backpropagation or some option teaches neural feed networks. Its greatest problem is that in most applications they aren't fast. The evaluation of optimum primary weights is one methodology for faster learning speed. A further practice is the multi-layered perceptron with forward ANN feed, which is an algorithm for weight removal. This algorithm creates the suitable topology spontaneously and avoids overfitting issues thereafter. Sometimes genetic tools are used to change cell weights and to look for the spatial layout of a neural network. Bayesian techniques are also used for neural network training. Verity of other techniques have been implemented recent efforts to improve the velocity and efficiency of ANN's teaching algorithms through changes in the structural design of the networks. These processes involve tapping insufficient weights or nodes and efficient algorithms when additional nodes are crucial [45] [46] [47].

In many sciences and engineering field, another well-known approach Radial Basis Function (RBF) networks rapidly applied in the field of machine learning. A three-layer feedback network comprises an RBF network. In this three-layer neuron, the hidden layer neuron uses the RBF or radial-basis function. The process of this type of learning is performed over two phases. Initially, cluster algorithms determine the cores and widths of the hidden layer. Single value decomposition (SVD) or the lowest mean squared (LMS) methods help to determine the weights of the hidden layer. A major challenge for the RBF network is with the challenge of selecting the appropriate quantity of fundamental features. The quality and over-simplification of RBF networks are controlled by the fundamental function. RBF networks with too limited base functions can not be sufficiently flexible for the training data. Usage of too many basis functions for generalization purpose produces noise in training data. Two diverse techniques are multilayer ANN and decision tree. The overall conclusions in symbolic learning approaches are i) neural networks are naturally capable to offer to learn in an incremental way than decision trees. Induction procedures of incremental decision tree outcome in recurrent tree rearrangement when the number of training data is lesser, with the tree organization growing as a pool of the data turn into bigger. ii) In terms of time consumed by training algorithm for the neural network is higher than that of the decision tree [48] [49] [50] [51].

1.2.3 Statistical learning algorithms

Statistical methodology is an implicit, probability-based model. Instead of classification, this method gives an opportunity for an event to fit within each category. In the machinery and statistical field, linear discriminatory (LDA) and the associated Fisher linear discriminant are very straightforward and common methods. For isolating subjects in two or more courses, the characteristics that consist of a linear mixture are superlative. LDA operates if capabilities and sizes are ongoing quantities ready for individual observation. When cases are categorically organized, Discriminant Correspondence Analysis is the appropriate method. Maximum entropy is a popular extra technique for approximating information probability allocations. The main value in the maximum entropy is that the distribution should vary so much as it is promising if everything is unknown. Categorized teaching information is used to create a number of procedural limitations which illustrate the class-specific allocation outlook. The symbolic symbols for statistic learning algorithms are the Bayesian networks [52] [53].

Naive Bayesian networks (NB) are another type of appropriate Bayesian networks. NB is directed acyclic graphs composed of only the undetected node (parents) and several recognized nodes (children) with a powerful statement of independence surrounded by child nodes in the parent's scenario. The guidance system (Naive Bays) is designed to fit these two probabilities, and also the higher likelihood provides more accuracy. The evaluation of the Bayes probability influenced by client intervention. It is specifically willing to be improperly obstructed by zero opportunities. The hypothesis of independence surrounded by child nodes is evidently generally wrong and therefore Bayes classifiers are commonly not as much accurate than other more refined learning algorithms such as Artificial Neural Networks. The Bayes model is easily autonomous, updated in several ways in order to enhance its effectiveness. Efforts to dim the hypothesis of independence are mainly based on the accumulation of additional borders to take account of the dependencies of certain features. In this situation, the network merely has the limit to combine an individual function with another function. In this scenario, the network has the limit to merely connect an individual function with another. The Bayesian semi-nautical classifier is another important attempt to prevent the hypothesis of independence. Attributes here are separated into clusters and each item, if only and only if in different clusters, is presumed to be autonomous of other items. The biggest advantage of the naive classification of Bayes is its computer time for practice. [27] [54] [55].

A Bayesian Network (BN) is used for chances and possibilities of relationships between features sets. In a BN structure, a directed acyclic graph (DAG) is denoted with S and the features X are establishing peer-to-peer communication in node S. The curves indicate occasional effects, while the lack of likely S arcs limits independence. Of course, a Bayesian network training assignment can be broken down into two subtasks: firstly, the DAG network training construct and then its parameters' intent. Tables set are encrypted with probabilistic parameters. Factors are dispersed in the indigenous condition of their families. Only these diagrams can replicate the network to determine the particular allocation independencies. There are two countries in the shared framework of the Bayesian networks: the established structure and the undetected framework. In the first state, a knowledgeable and presumed right is defined in the network's method. Learning the conditions of probability variables as quickly as the network composition is continuous. Tables (CPT) is generally explained by approximating a nearby exponential number of factors from the data supplied. In the network, every node has a connected individually to CPT that defines the restricted probability scattering of that node specified the dissimilar values of its parents. Bayesian Networks has the difficulty of finding a previously unacknowledged network. In N features problem, the number of possible hypotheses of the structure is higher than exponential in N [56] [57] [58].

1.2.4 Instance-based learning

Instance-based teaching is a fresh form of teaching under statistical methods. Lethargic training is regarded as instance-based teaching processes. This course takes time for the instruction or generalization to be completed. During the teaching stage, lethargic systems require less time than passionate teaching algorithms like decision-making bodies, neural networks and Bayes networks. However, during the classification phase, it requires more calculation time than passionate teaching algorithms. One of the finest first examples of teaching is the nearest neighbour (KNN) algorithm. k-Nearest Neighbor (KNN) is based on the conviction that in the neighbourhood the cases of a dataset with similar characteristics will be existing. If the instances have a grouped tag, then the tag value can be determined by finding the next neighbour class of an uncategorized instance. The KNN discovers the k cases closest to the query example by classifying the highest standard class label. The cases can be evaluated as points within a multidimensional example room in a broad spectrum. The whole place of cases within this room is not as significant as the comparative range between cases. By using a distance metric this comparative distance is defined. Ideally, the range measurement requires to decrease the distance between two similarly categorized cases while improving the distance between different categories. In many true areas, the force of KNN is created. The selection of k is effective in the KNN algorithm. There are two explanations why the nearest neighbour classifier classifies an example inaccurately. In the first place, the noisy cases succeed in the mass ballot when noise is present in the neighbourhood of the request example, which results in an inadequate class. A superior k might solve this problem in this situation. Second, when the class or class region is too small to suit the class that surrounds the piece, the common ballot will succeed. A smaller k might solve this problem. As we pointed out, their great calculation time for classification is the main drawback to instance-based classification devices. A primary problem in several apps is to solve the current input characteristics to demonstrate by selecting the function, as this can improve classification precision and decrease the required time for computing. In addition, choosing an extra suitable data set range measure can enhance the accuracy of instance-based classifiers [59] [60] [61] [62].

1.2.5 Support Vector Machines

SVM is one of the contemporary machine learning techniques. SVM's revolve around the „margin" notion— both sides of a hyperplane dividing two different categories of information. Making complete use of the margin is acknowledged to reduce the upper limit on its predictable inaccuracies with regard to the feasible range between the divisive hyperplanes. Informally, an optimum dividing hyperplane can be looked for, if it promises linear isolation of two classes, by reducing the adjustment of the hyperplane. In the circumstance of linearly divisible data, when the optimal splitting hyperplane is searched, data points that fib on its margin are acknowledged as support vector points and the resolution is characterized as a linear grouping of only these points. Extra data points are disregarded. Thus, the process complexity of an SVM is natural by features number seen in the training data. Therefore, SVMs are satisfactory and appropriate to deal with learning responsibilities where the features set is huge regarding the training instances set [63] [64] [65] [66].

1.2.6 Unsupervised Learning

Unlike supervised learning, in unsupervised learning, there is no categorized training data to study from and no forecast to be prepared. In unsupervised learning, the main goal is to take a dataset as input and attempt to search patterns or normal groupings inside the data elements of records. Hence, unsupervised learning is regularly named as a descriptive model and the method of unsupervised learning is denoted as pattern finding or knowledge finding. One thoughtful application of unsupervised learning is splitting up of customer.

Clustering is the key concept of unsupervised learning. It proposes to make cluster or grouping related entities or objects together. Therefore, objects fitting to the similar cluster are relative to everyone while objects fitting to the dissimilar cluster relatively different. Hence, the objective of clustering to determine the essential grouping of uncategorized data from clusters. Dissimilar measures of connection can be useful for clustering. For measuring similarity, the distance between the objects is generally accepted. Therefore, we can say that two data objects are in the same cluster if the distance between the two clusters is very small or ignorable. In the same way, if the distance in the middle of the data objects is large, then the objects do not usually fit the same cluster. This type of learning is also recognized as distance-based clustering. Except clustering of data, a different alternative supervised learning is association analysis. In association analysis, the association in the middle of the data features is identified [67] [68] [69].

1.2.7 Reinforcement Learning

In reinforcement learning, a device learns from the performance on its individual to succeed the specified goals. This learning category is associated with self-learning by a human. A common example is a baby getting the knowledge to cross the hurdle. Without having any knowledge, He makes collision on to hurdles and falls down several times. He is getting knowledge whenever he crosses over the hurdles. He appears in a similar issue while learning to cycle riding as a child or car driving as a beginner. Not everything is trained by experts. Many things are essential to knowledge only from faults or mistakes done in previous. We incline to start a specification on things that we should do and do not construct on past knowledge. Self-driving vehicles are the one existing example of reinforcement learning. The life-threatening information that is essential to be careful are speed and speed limit of vehicles in dissimilar road divisions, traffic situation, road conditions, weather circumstances, etc. [70] [71] [72].

1.3 Discussion and Analysis

Supervised machine learning methods can be applied in many areas. SVMs and neural networks appear to work faster in a multi-dimension dataset with a series of features. In communicating with separate or categorical features, logic-based systems work faster. Neural network and SVM models need a big dataset, in which a tiny dataset is necessary for NB. K-NN is typical of irrelevant characteristics. Training in the neural network may be very inefficient although there are insignificant features. Logic-specific algorithms all of the neural networks and SVMs are regarded as very simple to understand. k-NN is also regarded to be very poorly interpreted because it is not easy to read an unstructured training instance.

In general, controlled teaching enables the label concept to be very particular. That is to say, the algorithm can be taught to differentiate between distinct categories where an optimal choice limit is established. You can determine the number of courses needed. The input information is well recognized and marked with this teaching methodology. Compared to the outcomes generated through the unattended master training techniques, the findings generated by the monitored methods are more precise and reliable. This is primarily due to the well recognized and marked input information in the monitoring algorithm. This is a significant distinction between controlled and uncontrolled teaching. The responses in the analysis and algorithm production are probably known because every class used is recognized.

In comparison with the unsupervised method, supervised learning can nevertheless be complex. The main reason for this is the very well-understood and labelled inputs to monitored education. Nobody is needed to comprehend and then mark information entries, unlike monitored algorithms, in uncontrolled teaching. Unmonitored teaching is, therefore, less complicated and why many individuals favour unmonitored methods. Supervised learning is not in real-time while unattended learning is in real-time. This also represents a significant distinction between monitored and unattended teaching. All the input information should be analyzed and marked in the presence of students in uncontrolled teaching. This helps to comprehend very diverse designs of raw information processing and processing. Moreover, the retrieval of information from a laptop is often simpler than that of labelled information requiring action by individuals. Off-line assessment is used for supervised machine learning. A lot of training moment is required for computation. When the data input is large and increasingly large, the rules are uncertain in the labels to be predefined. This can be a difficult task.

In most cases, interactive software systems or applications are used in supervised learning methodologies. In Artificial Intelligence, where human communication prevails, reinforcement learning promotes and operates better. Supervised learning is a region of machine learning where the assessment of a general formula for a software system can be done only with sample information for the teaching of the scheme through the use of the training data or the examples provided to the model. In order to attain a behavioural event, reinforcement learning has a training officer who interacts with the setting to look at the fundamental behaviour of the human structure. Control theory, study, gaming theory, information theory and so on are the purposes. The applications of supervised and reinforcement learning differ on the purpose or goal of a software system. Both Supervised Learning and Reinforcement Learning have huge advantages in the area of their applications in computer science. The development of different new algorithms causes more development and improvement of performance and growth of machine learning that will result in sophisticated learning methods in supervised learning as well as reinforcement learning.

Reinforcement Learning has an input-to-output tracing framework which directs the system. Unsupervised Learning, however, does not contain such characteristics. The fundamental challenge of finding the patterns instead of mapping to achieve the ultimate objective lies in Unsupervised Learning. For instance, if a successful press update is recommended for a consumer, a Reinforcement Learning algorithm looks at receiving frequent feedback from the customer concerned, and from the knowledge graph of all media posts are presented to the individual. Rather, an Unsupervised Learning algorithm will attempt to look at many other papers, similar to the one read by the person, and suggest something that fits the preferences of the user.

1.4 Conclusion

The usage of the supervised, unsupervised and reinforcement learning are discussed with their advantages and disadvantages. The survey of different techniques of learning process indicates that the hybridisation of learning techniques is beneficial if and only if the efficiency is improved in a modified version of the algorithm. This can be surely possible if the hybridisation is performed in such a way that the disadvantage of one algorithm is treated by the advantages of another algorithm. It is always not so important to justify a classifier for its highest accuracy but it is better to look into three factors of a learning algorithm. The entire storing capacity be determined by the size of the individual module classifier itself and the size of the group (total number of classifiers exists in the group). The second disadvantages are bigger computation time because an input query can classify appropriately if all component classifiers (in its place of a particular classifier) must be processed. The last disadvantages are reduced by unambiguousness. Here, the ideas are thoroughly discussed in order to ensure that readers are constantly conscious of the advancement towards an artificially intelligent culture. When considering the machine learning algorithm ranking, the primary study is not whether the learning algorithm is stronger than the others, but under which conditions a given method can overcome others considerably on a particular implementation issue. Supervised, unsupervised and reinforcement machine learning are a description of ways in which machines or algorithms suffer on a data set. Computation cost is maximum for modern learning algorithms and memory is occupied by the data that is obviously inappropriate for several reasonable problems. An orthogonal methodology is splitting the data, escaping the necessity to execute algorithms on enormous datasets. The idea of distributive machine learning includes the division into subsets of the dataset, the simultaneous learning from these sub-sets and the combination of the results. Over moment, supervised, unchecked and strengthened teaching leads to the future of computers that are supposed to be brilliant and will help people to do everyday stuff.

References

1. Holte, R, 1989. "Alternative information structures in incremental learning systems, In: Machine and Human Learning – Advances in European Research, Kodrato, Y and Hutchinson, A (eds.). 121–142. Kogan Page.
2. Fielding, A., 1999. Machine Learning Methods for Ecological Applications. Springer Science & Business Media.
3. Y. Bengio , Learning deep architectures for AI, Found. Trends Mach. Learn. 2 (1) (2009) 1–127.
4. David M. Dutton, Gerard V. Conroy, A review of machine learning. The Knowledge Engineering Review, Vol. 12:4, 1996, 341–367
5. Bishop, C.M., 2006. Pattern Recognition and Machine Learning. Springer, New York, USA.
6. I.H. Witten, E. Frank, Data Mining: Practical Machine Learning Tools and Techniques, Morgan Kaufmann, 2005.
7. S. Russell, P. Norvig, Artificial intelligence: A Modern Approach, Prentice Hall, Upper Saddle River, 2009.
8. Amara, J., Bouaziz, B., Algergawy, A., 2017. A Deep Learning-Based Approach for Banana Leaf Diseases Classification. BTW workshop, Stuttgart, pp. 79–88.
9. Deng, L., Yu, D., 2014. Deep learning: methods and applications. Found. Trends Signal Process. 7 (3–4), 197–387.
10. Carbonell, J, Michalski, R and Mitchell, T, 1983. "An overview of machine learning", In: Michalski, R, Carbonell, J and Mitchell, T, eds., Machine Learning: An AI Approach, Morgan-Kaufmann.
11. Spears, Wand De Jong, K, 1990. "Using genetic algorithms for supervised concept learning". Proceedings 2nd International IEEE Conference on Tools for AI 335–341. IEEE Press.
12. Wang, W and Chen, J, 1991. "Learning by discovering problem solving heuristics through experience". IEEE Transactions on Knowledge and Data Engineering 3(4) December, 415±419.
13. Weiss, S and Kulikowski, C, 1991. Computer Systems That Learn – Classification and Prediction Methods from Statistics, Neural Nets, Machine Learning and Expert Systems Morgan-Kaufmann.

14. Winkel bauer, L and Fedra, K, 1991. "ALEX: Automatic Learning in Expert Systems". Proceedings 7th Conference on Artificial Intelligence Applications 59±62. IEEE Press.

15. Batista, G., & Monard, M.C., (2003), An Analysis of Four Missing Data Treatment Methods for Supervised Learning, Applied Artificial Intelligence, vol. 17, pp.519–533.

16. Quinlan, J, 1986b. "The effect of noise on concept learning". In: Machine Learning: An AI Approach vol 2, Michalski, R, Carbonell, J and Mitchell, T (eds.). Morgan-Kaufmann.

17. J. Platt, Sequential minimal optimization: a fast algorithm for training support vector machines, Microsoft Research, 1998. Technical report msr-tr-98–14.

18. A. Fallahi , S. Jafari , An expert system for detection of breast cancer using data pre-processing and Bayesian network, Int. J. Adv. Sci. Technol. 34 (2011) 65–70.

19. Jia, Y., Shelhamer, E., Donahue, J., Karayev, S., Long, J., Girshick, R., Darrell, T., 2014. Caffe: Convolutional architecture for fast feature embedding. In: Proceedings of the 22nd International Conference on Multimedia. ACM, Orlando, FL, USA, pp. 675–678.

20. Chen, Y., Lin, Z., Zhao, X., Wang, G., Gu, Y., 2014. Deep learning-based classification of hyperspectral data. IEEE J. Sel. Top. Appl. Earth Obs. Remote Sens. 7 (6), 2094–2107.

21. Lantz, B., 2013. Machine Learning with R. Packt Publishing Ltd.

22. Y. Freund, R.E. Schapire , Experiments with a new boosting algorithm, in: Proceedings of International Conference on Machine Learning, vol. 96, 1996, pp. 148–156.

23. M. Kubat, R.C. Holte, S. Matwin, Machine learning for the detection of oil spills in satellite radar images, Mach. Learn. 30 (2) (1998) 195–215, doi: 10.1023/A: 1007452223027.

24. G. Batista , R.C. Prati , M.C. Monard , A study of the behavior of several methods for balancing machine learning training data, ACM SIGKDD Explor. Newsl. 6 (1) (2004) 20–29.

25. H. Guo , H.L. Viktor , Learning from imbalanced data sets with boosting and data generation: the databoost-IM approach, ACM SIGKDD Explor. Newsl. 6 (1) (2004) 30–39.

26. Y. Liu , N.V. Chawla , M.P. Harper , E. Shriberg , A. Stolcke , A study in machine learning from imbalanced data for sentence boundary detection in speech, Comput. Speech Lang. 20 (4) (2006) 46 8–4 94.

27. S. B. Kotsiantis. Supervised Machine Learning: A Review of Classification Techniques. Informatica 31 (2007) 249–268.

28. Safavian, S and Landgrebe, D, 1991. "A survey of decision tree classifier methodology". IEEE Transactions on Systems, Man and Cybernetics 21(3) May/June, 660–674.

29. Baik, S. Bala, J. (2004), A Decision Tree Algorithm for Distributed Data Mining: Towards Network Intrusion Detection, Lecture Notes in Computer Science, Volume 3046, Pages 206–212.
30. Breslow, L. A. & Aha, D. W. (1997). Simplifying decision trees: A survey. Knowledge Engineering Review 12: 1–40.
31. Elomaa T. (1999). The biases of decision tree pruning strategies. Lecture Notes in Computer Science 1642. Springer, pp. 63–74.
32. Scha er, C, 1991. "When does overfitting decrease prediction accuracy in induced decision trees and rule sets".
33. Kuhn, M., Weston, S., Coulter, N., Quinlan, R., 2015. C50: C5.0 Decision Trees and Rule-Based Models. ? R Package Version 0.1. pp. 0–24. https://CRAN.R-project.org/package=C50.
34. Donald, JH, 1994. "Rule induction – Machine learning techniques". Computing and Control Engineering Journal 5(5) October, 249±255.
35. Cohen, W. (1995), Fast Effective Rule Induction. In Proceedings of ICML-95, 115–123.
36. Furnkranz, J. (1997). Pruning algorithms for rule learning. Machine Learning 27: 139–171.
37. Freund, Y. & Schapire, R. (1999), Large Margin Classification Using the Perceptron Algorithm, Machine Learning 37: 277–296.
38. Peter Auera, Harald Burgsteinerb Wolfgang Maassc. A learning rule for very simple universal approximators consisting of a single layer of perceptrons. Neural Networks Volume 21, Issue 5, June 2008, Pages 786–795
39. Steve Gallant. Perceptron-based learning algorithms. IEEE Transactions on Neural Networks · Volume 1, No. 2, June 1990
40. Jiexiong Tang ; Chenwei Deng ; Guang-Bin Huang. Extreme Learning Machine for Multilayer Perceptron. IEEE Transactions on Neural Networks and Learning Systems. Volume: 27, Issue: 4 , April 2016.
41. G. Huang, S. Song, J. N. D. Gupta, C. Wu, „Semi-supervised and unsupervised extreme learning machines", IEEE Trans. Cybern., vol. 44, no. 12, pp. 2405–2417, Dec. 2014.
42. M.W Gardnera, S.R. Dorlinga. Artificial neural networks (the multilayer perceptron)—a review of applications in the atmospheric sciences. Atmospheric Environment Volume 32, Issues 14–15, 1 August 1998, Pages 2627–2636
43. G. E. Hinton, R. R. Salakhutdinov, „Reducing the dimensionality of data with neural networks", Science, vol. 313, no. 5786, pp. 504–507, 2006.
44. N.-Y. Liang, G.-B. Huang, P. Saratchandran, N. Sundararajan, „A fast and accurate online sequential learning algorithm for feedforward networks", IEEE Trans. Neural Netw., vol. 17, no. 6, pp. 1411–1423, Nov. 2006.
45. Lin Wanga, Yi Zenga, Tao Chen. Back propagation neural network with adaptive differential evolution algorithm for time series forecasting. Expert Systems with Applications. Volume 42, Issue 2, 1 February 2015, Pages 855–863.

46. Roberthecht-Nielsen. Theory of the Backpropagation Neural Network. Neural Networks for Perception Computation, Learning, and Architectures 1992, Pages 65–93.
47. Cios, K and Liu, N, 1992. "A machine learning method for generation of a neural network architecture: a continuous ID3 algorithm". IEEE Transactions on Neural Networks 3(2) 280–290.
48. J. Park. I. W. Sandberg, Universal Approximation Using Radial-Basis-Function Networks. Neural Computation. Volume: 3 , Issue: 2 , June 1991.
49. Hossam Faris, Ibrahim Aljarah, Seyedali Mirjalili. Evolving Radial Basis Function Networks Using Moth–Flame Optimizer. Handbook of Neural Computation 2017, Pages 537–550
50. Robert, J., Howlett L.C.J. (2001), Radial Basis Function Networks 2: New Advances in Design.
51. C. Harpham author, C. W. Dawson, M. R. Brown. A review of genetic algorithms applied to training radial basis function networks. Neural Computing & Applications September 2004, Volume 13, Issue 3, pp 193–201
52. Vladimir N. Vapnik. An Overview of Statistical Learning Theory. IEEE Transactions on Neural Networks, Vol. 10, No. 5, September 1999
53. Iniesta, R., Stahl, D., & McGuffin, P. (2016). Machine learning, statistical learning and the future of biological research in psychiatry. Psychological Medicine, 46(12), 2455–2465. doi:10.1017/S0033291716001367
54. Langarizadeh M, Moghbeli F. Applying Naive Bayesian Networks to Disease Prediction: a Systematic Review. Acta Inform Med. 2016;24(5):364–369. doi:10.5455/aim.2016.24.364–369
55. Sharma RK, Sugumaran V, Kumar H, Amarnath M. A comparative study of naïve Bayes classifier and Bayes net classifier for fault diagnosis of roller bearing using sound signal. International Journal of Decision Support Systems. 2015 Jan 1;1(1):115–29.
56. Nir Friedman, Dan Geiger. Moises Goldszmidt. Bayesian Network Classifiers. Machine Learning November 1997, Volume 29, Issue 2–3, pp 131–163
57. Jacinto Arias, Jose A. Gamez, Jose M.Puerta. Learning distributed discrete Bayesian Network Classifiers under MapReduce with Apache Spark. Knowledge-Based Systems Volume 117, 1 February 2017, Pages 16–26
58. I. Cohen N. Sebe ; F.G. Gozman ; M.C. Cirelo ; T.S. Huang. Learning Bayesian network classifiers for facial expression recognition both labeled and unlabeled data. IEEE Computer Society Conference on Computer Vision and Pattern Recognition, 2003. Proceedings.
59. David W. Aha, Dennis Kibler, Marc K. Albert. Instance-based learning algorithms. Machine Learning January 1991, Volume 6, Issue 1, pp 37–66
60. Jordi L. Vermeulen , Arne Hillebrand, Roland Geraerts. A comparative study of k-nearest neighbour techniques in crowd simulation. Computer Animation and Virtual Worlds. Volume28, Issue3–4 May/August 2017.

61. Li-Yu Hu, Min-Wei Huang, Shih-Wen Ke and Chih-Fong Tsai. The distance function effect on k-nearest neighbor classification for medical datasets. Hu et al. SpringerPlus (2016) 5:1304
62. Bartosz A. Nowak, Robert K. Nowicki, Marcin Woźniak, Christian Napoli. Multi-class nearest Neighbour Classifier for Incomplete Data Handling. International Conference on Artificial Intelligence and Soft Computing ICAISC 2015: Artificial Intelligence and Soft Computing pp 469–480
63. Christopher J.C. Burges. A Tutorial on Support Vector Machines for Pattern Recognition. Data Mining and Knowledge Discovery June 1998, Volume 2, Issue 2, pp 121–167
64. Wu, W., Li, A.D., He, X.H., Ma, R., Liu, H.B., Lv, J.K., 2018. A comparison of support vector machines, artificial neural network and classification tree for identifying soil texture classes in southwest China. Comput. Electron. Agric. 144, 86–93.
65. Platt, Sequential minimal optimization: a fast algorithm for training support vector machines, Microsoft Research, 1998. Technical report msr-tr-98–14.
66. Cristianini, N. & Shawe-Taylor, J. (2000). An Introduction to Support Vector Machines and Other Kernel-Based Learning Methods. Cambridge University Press, Cambridge.
67. H.B. Barlow. Unsupervised Learning. Ncurd Coiiipiftntioi~ 1, 295–311 (1989)
68. Terrence J. Sejnowski, Laurenz Wiskott. Slow Feature Analysis: Unsupervised Learning of Invariances. Neural Computation 14, 715–770 (2002).
69. Hastie T., Tibshirani R., Friedman J. (2009) Unsupervised Learning. In: The Elements of Statistical Learning. Springer Series in Statistics. Springer, New York, pp 485–585.
70. Leslie Pack Kaelbling, MichaelL Littman, Andrew W.Moore. Reinforcement Learning: A Survey Journal of Artificial Intelligence Research 4 (1996) 237–285.
71. Barto, A. G. & Sutton, R. (1997). Introduction to Reinforcement Learning. MIT Press.
72. Jens Kober, J. Andrew Bagnell, Jan Peters. Reinforcement learning in robotics: A survey, The International Journal of Robotics Research. Vol 32, Issue 11, 2013

Shashwati Mishra, and Mrutyunjaya Panda

2 An Analysis on Non-linear Dimension Reduction Techniques

Abstract: Rapid use of social media for communication and information sharing, use of world wide web as a huge information repository, advancement in data and image capturing techniques have drastically increased the volume and size of the data. Analyzing, processing and searching such a huge volume of data is a complex task. This also reduces the accuracy as well as diminishes performance. To overcome these problems dimension reduction techniques are used to reduce the size of data and features without the loss of valuable information. Dimension of features can be reduced by projecting the features to a low dimensional space or using feature selection or feature encoding techniques. These techniques can be linear or non-linear depending on the relationship between data in higher and lower dimensions. In this chapter, a detailed analysis has been performed on some of the non-linear dimensionality reduction techniques and their applications.

Keywords: Linear dimension reduction, Non-linear dimension reduction, k-PCA, Hessian LLE, t-SNE, Diffusion maps, Generalized Discriminant Analysis

2.1 Introduction

The developments in data collection methods, data transmission techniques, sensing devices, storage devices, image capturing devices, communication technology have tremendously increased the size and volume of data. But this increase in data size has also become a barrier in faster and better analysis of data. Dimension reduction techniques are used to convert high dimensional data to low dimension by preserving the valuable information present in the high dimensional data. Dimension reduction techniques are used in machine learning to reduce the dimension of extracted features which helps in reducing the complexity of classification, analysis and processing activities.

Dimension reduction techniques reduce the size of original data by removing less relevant attributes. This helps in reducing the storage space required to store the data and helps in easy visualization of the data. Better visualization makes the process of analysis easier. Reduced volume of data also makes the computation easier and faster. Some irrelevant information may also be removed which helps in reducing noise and increasing the accuracy of computation. The data processing operations become more complex due to the presence of high dimensional data. Reduced dimension of data helps in easy and faster classification, clustering, analysis, prediction of data.

https://doi.org/10.1515/9783110610987-004

Now-a-days advanced data capturing technologies, communication technologies, storage technologies have increased the volume of data generated. In comparison to earlier days dependency on computers has also increased to get faster and error free results.

Another advantage of dimension reduction is it helps in removing the redundant and noisy information from the data. Some data values may be irrelevant and noisy, some may be redundant. Such values unnecessarily increase the volume of the data and also the complexity of processing operations. Dimensionality reduction techniques remove such information by selecting and extracting important features.

Due to the presence of redundant and irrelevant data successfully trained neural networks do not give good result for test data. Extraction of relevant and important information from redundant and irrelevant information is very difficult and increases the learning time of the training phase [1].

The accuracy of dimension reduction depends on the type of data and presence of noise in the data. Extracting relevant information from the huge volume of data without loss of information is also a challenging task.

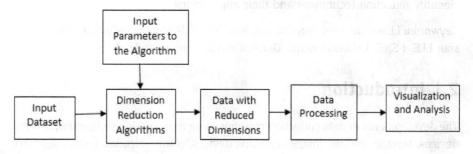

Fig. 2.1: Dimension reduction steps.

2.1.1 Categorization of Dimension Reduction Techniques

Dimensionality reduction techniques can be categorized as convex or non-convex depending on the presence or absence of local optima. Non-convex techniques are applied where there is the presence of local optima otherwise convex techniques are used. Convex methods can further be classified as full spectral and sparse spectral techniques depending on the type of matrix used for eigen decomposition. Full spectral methods use full matrix and sparse spectral method use sparse matrix. PCA (Principal Component Analysis), kernel PCA, Isomap, Diffusion maps, Maximum Variance Unfolding are some examples of full spectral techniques. These methods use a full matrix which contains the similarity value between data points or covariance value between dimensions. Sparse techniques which use sparse matrix try to preserve the local structure of the data. LLE (Local Linear Embedding), Hessian LLE, Laplacian Eigenmaps, LTSA (Local Tangent Space Analysis) are some examples of sparse spectral techniques. Sammon mapping, LLC (Locally Linear Coordination), multilayer autoencoders, manifold charting are some non-convex techniques of dimension reduction. L. Maaten et al. performed a detailed analysis on these techniques of dimension reduction [3].

Dimensions can be reduced either by selecting the desired number of features or by projecting the high dimensional features to low dimensional features or by encoding the features. Feature selection techniques select the most relevant data values to get better prediction accuracy with less computation time. Selection of appropriate features also helps in reducing noise and irrelevant information. Feature selection methods can be of two types filter methods and wrappers methods. Projection or transformation of features from high to low dimensional space is obtained by generating a small set from the input set of variables. This small set represents the original input information without any redundancy, noise and irrelevant information. Feature transformation can be linear or non-linear transformation. Feature encoding techniques are used to encode the data in compact forms for better storing, retrieval and analysis. Quantization, hashing techniques are used to encode the features [2]. A tree structure view of the different ways of classifying the dimension reduction techniques is given in Figure 2.2.

Non-linear techniques are popularly used for better understanding, analysis and visualization of complex data and information. Linear dimensionality reduction techniques are suitable for linear data which are the oldest approach of dimension reduction. Gradually to reduce the dimension of non-linear data non-linear dimensionality reduction methods were developed that gives better result than linear methods. Manifold learning techniques were emerged along with the concept of geodesic distance [4]

Section 2 of the chapter explains about five non-linear dimension reduction techniques and research works on dimension reduction techniques are analyzed in section 3. Section 4 discusses about the experimental observations followed by conclusion section.

2.2 Non-linear Dimension Reduction Techniques

The techniques of dimensionality reduction may include Principal Component Analysis (PCA), Linear Discriminant Analysis (LDA), Factor Analysis (FA), Locally Linear Embedding (LLE), Non-negative Matrix Factorization (NMF), Independent Component Analysis (ICA), Generalized Discriminant Analysis (GDA), Latent Semantic Analysis (LSA), Multi-dimensional Scaling (MDS) etc. Deep learning based methods like Autoencoders and Restricted Boltzmann Machines (RBMs) are also used to reduce the dimension of extracted features [2]. The dimensionality reduction techniques can be linear or non-linear depending on the relationship between the data in higher dimension and lower dimension.

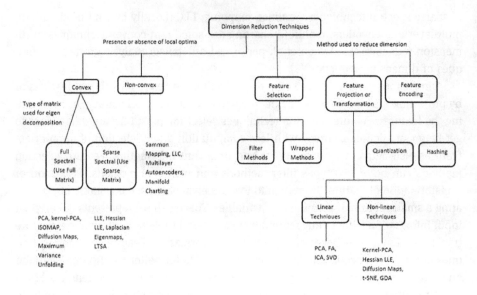

Fig. 2.2: Classification of dimension reduction techniques.

2.2.1 Kernel PCA

PCA is one of the popular dimensionality reduction methods, but it cannot classify data which are non-linear in nature. So kernel method is applied to convert non-linear data to linear format on which PCA can be applied. Instead of using the feature mapping function to generate the kernel map some popular kernel functions like Gaussian and polynomial function are used to generate the feature space. Selecting a valid kernel function is very important for kernel substitution. Kernel functions can be constructed directly or can also be built from simple kernels. Another method first finds a feature space mapping and then using this feature space tries to get the kernel where kernel function is defined as in equation (2.1).

$$K(x_m, x_n) = \phi(x_m)^T \phi(x_n) = \sum_{i=1}^{k} \phi_i(x_m)\phi_i(x_n) \tag{2.1}$$

where, $\phi_i(x_m)$ are the basis functions. Kernel PCA is a non-linear dimensionality reduction technique which applies this kernel substitution technique to PCA (Principal Component Analysis). Suppose the data set consists of $\{x_m\}$ for $m = 1, 2, \ldots\ldots\ldots, M$ observations $x_1, x_2, \ldots\ldots\ldots, x_M$ in a D dimensional space. Assume that each data point x_m is projected into a feature space with a non-linear transformation. Then the projected point after applying the non-linear transformation will be $\phi(x_m)$ [5]. Suppose the mean of projected data points is zero. Mathematically,

$$\mu = 1M \sum_{m=1}^{M} \phi(x_m) = 0 \tag{2.2}$$

The covariance matrix will be, $C = 1M \sum\limits_{m=1}^{M} \phi(x_m)\phi(x_m)^T$

$$\text{(2.3)}$$

Assuming the dimension of feature space as$N$, the eigenvector equation becomes

$$Cv_i = \lambda_i v_i \text{ for } i = 1, 2, \ldots\ldots, N \tag{2.4}$$

Using equation (2.3) in equation (2.4) $1M \sum\limits_{m=1}^{M} \phi(x_m)\phi(x_m)^T v_i = \lambda_i v_i$

$$\text{(2.5)}$$

$$\Rightarrow v_i = 1M\lambda_i \sum\limits_{m=1}^{M} \phi(x_m)\phi(x_m)^T v_i$$

$$= 1M\lambda_i \sum\limits_{m=1}^{M} \left(\phi(x_m)v_i\right)\phi(x_m)^T \tag{2.6}$$

For $(\lambda_i > 0)$, $v_i = \sum\limits_{m=1}^{M} \alpha_{im}\phi(x_m)$

$$\text{(2.7)}$$

Equation (2.7) proves that the eigenvectors can be represented as a linear combination of feature points. Using equation (2.7) in (2.5)

$$1M \sum\limits_{m=1}^{M} \phi(x_m)\phi(x_m)^T \sum\limits_{n=1}^{M} \alpha_{in}\phi(x_n) = \lambda_i \sum\limits_{m=1}^{M} \alpha_{im}\phi(x_m) \tag{2.8}$$

Multiplying both sides of equation (2.8) by$\phi(x_j)^T$, we get

$$1M \sum\limits_{m=1}^{M} \phi(x_m)\phi(x_j)^T \sum\limits_{n=1}^{M} \alpha_{in}\phi(x_n)\phi(x_m)^T = \lambda_i \sum\limits_{m=1}^{M} \alpha_{im}\phi(x_m)\phi(x_j)^T$$

$$\Rightarrow 1M \sum\limits_{m=1}^{M} K(x_j, x_m) \sum\limits_{n=1}^{M} \alpha_{in}K(x_m, x_n) = \lambda_i \sum\limits_{m=1}^{M} \alpha_{im}K(x_j, x_m) \tag{2.9}$$

Equation (2.9) can be written in matrix form as

$$K^2\alpha_i = \lambda_i MK\alpha_i \tag{2.10}$$

where, α_i is a column vector containing elements α_{mi}, for $m = 1, 2, \ldots, M$. Removing factor K from both sides

$$K\alpha_i = \lambda_i M\alpha_i \tag{2.11}$$

Equation (2.11) can be solved to obtain solutions for α_i. Normalizing the eigenvectors in feature space helps in getting the normalization condition forα_i[5].

$$v_i^T v_i = 1 \Rightarrow \sum\limits_{m=1}^{M} \alpha_{im}\phi(x_m)^T \sum\limits_{n=1}^{M} \alpha_{in}\phi(x_n) = 1 \text{(Using equation (2.7))}$$

$$\Rightarrow \sum_{m=1}^{M} \sum_{n=1}^{M} \alpha_{im}\alpha_{in}\phi(x_m)^T\phi(x_n) = 1$$

$$\Rightarrow \alpha_i^T K \alpha_i = 1$$

$$\Rightarrow \lambda_i M \alpha_i^T \alpha_i = 1 , \forall i \text{ [Substituting the value of } K \text{ from equation (2.11)]}$$ (2.12)

The principal component projection of any point x on eigenvector i can be represented in terms of kernel function as:

$$y_i(x) = \phi(x)^T v_i = \phi(x)^T \sum_{m=1}^{M} \alpha_{im}\phi(x_m)$$

$$= \sum_{m=1}^{M} \alpha_{im}\phi(x)^T\phi(x_m)$$

$$\Rightarrow y_i(x) = \sum_{m=1}^{M} \alpha_{im}K(x, x_m)$$ (2.13)

The projected data points may not have zero mean. The projected features can be centralized using equation (2.14).

$$\tilde{\phi}(x_m) = \phi(x_m) - 1M \sum_{l=1}^{M} \phi(x_l)$$ (2.14)

So the corresponding kernel becomes:

$$\tilde{K}_{mn} = \tilde{\phi}(x_m)^T\tilde{\phi}(x_n)$$

$$= [\phi(x_m) - 1M \sum_{l=1}^{M} \phi(x_l)]^T[\phi(x_n) - 1M \sum_{l=1}^{M} \phi(x_l)]$$

$$= \phi(x_m)^T\phi(x_n) - 1M \sum_{l=1}^{M} \phi(x_m)^T\phi(x_l) - 1M \sum_{l=1}^{M} \phi(x_l)^T\phi(x_n) + 1M^2 \sum_{l=1}^{M}\sum_{j=1}^{M} \phi(x_l)^T\phi(x_j)$$

$$= K(x_m, x_n) - 1M \sum_{l=1}^{M} K(x_m, x_l) - 1M \sum_{l=1}^{M} (x_l, x_n) + 1M^2 \sum_{l=1}^{M}\sum_{j=1}^{M} (x_l, x_j)$$

$$\Rightarrow \tilde{K} = K - 1_M K - K 1_M + 1_M K 1_M$$ (2.15)

Equation (2.15) is in matrix form where 1_M represents $M \times M$ matrix where each element has value 1/M. [5]

2.2.2 Hessian LLE (Locally Linear Embedding)

The dimensionality can be reduced using Hessian LLE if the manifold is connected. The Hessian LLE algorithm is a modification of LLE and the theoretical framework can be obtained from Laplacian eigenmap by using Hessian operator instead of Laplacian [6, 7]. Suppose x_i for $i = 1, 2, ..., N$ are the set of N points lying in a D dimensional space. To obtain the low dimensional representation of data Hessian LLE algorithm follows the following steps:

a) Neighbor identification
This step finds k-nearest neighbors of each data point x_i on the basis of Euclidean distance. Suppose Neg_i for $i = 1, 2, ..., N$ represents the neighborhood of point x_i. Each neighborhood contains k points, that means, $Neg_i = [x_{i_1}, x_{i_2}, ...x_{i_k}]$ is the neighborhood of x_i.

b) Tangent coordinates calculation
Singular Value Decomposition (SVD) is applied on the points in the neighborhood of x_i to find the tangent coordinates. For neighborhood of x_i it gives $U^i S_i (V^i)^T$. The tangent coordinates of Neg_i are represented by the first d columns of V^i.

c) Hessian estimator generation
Suppose H_i is the matrix and $f_j = (f(x_i))$. v^i vector represents the entries obtained from f by the extraction of entries corresponding to points in Neg_i. The matrix $H^i v^i$ gives an approximation of the Hessian matrix.

d) Quadratic form creation
Find the symmetric matrix

$$\bar{H}_{i,j} = \sum_p \sum_q ((H^p)_{q,i}(H^p)_{q,j})$$ (2.16)

The matrix $\bar{H}$ gives an approximation of the continuous operator H.

e) Low dimensional embedding
Eigen analysis is performed and eigenvectors corresponding to d smallest eigenvalues are chosen for low dimensional embedding [6, 7, 8].

2.2.3 Diffusion Maps

Diffusion maps which were introduced by R.R. Coifman, S. Lafon [9] is a non-linear dimensionality reduction technique and tries to obtain a global representation by integrating local geometry at different scales. The embedding is done in such a way that the Euclidean distance between the embed points gives an approximate representation of the diffusion distance of the features in the feature space [10].

Diffusion map algorithm involves the following steps:

a) Kernel matrix creation by random walk

The probability of moving to a closer point is more than moving to a distant point. The concept of connectivity has come from the idea that if probability of moving between the points is more than others then the points are nearer to each other (connected). So, for any two points a and b connectivity is the probability of moving from point a to point b and can be represented as [10]:

$$connectivity(a, b) = p(a, b) \tag{2.17}$$

Suppose dk is the kernel used to represent the similarity between the points of high dimensional dataset X. Since a kernel represents the local geometry by considering a specific characteristic of the data, the choice of kernel varies depending on the type of application [9]. The selection of neighborhood size depends on the nature of data like sparse, dense distribution of data points. Connectivity can be expressed in terms of diffusion kernel as:

$$connectivity(a, b) \propto dk(a, b) \tag{2.18}$$

The diffusion kernel $dk(a, b) \geq 0$ and is symmetric that means $dk(a, b) = dk(b, a)$. Assume that μ represents the distribution of the data points. Consider the local measure of the volume (or degree of the graph)

$$d(x) = \int_X dk(a, b) \, d\mu(y) \tag{2.19}$$

and $p(a, b) = dk(a, b)d(x) \tag{2.20}$

From equation (2.19) and equation (2.20), $\int_X p(a, b)d\mu(y) = 1 \tag{2.21}$

Suppose DM is the diffusion matrix whose entries represent the connectivity between the data points. Let

$$DM = \begin{bmatrix} dm_{aa} & dm_{ab} \\ dm_{ba} & dm_{bb} \end{bmatrix},$$

where, each entry dm_{ij} represents the connectivity between data points i and j which is nothing but the probability of moving from i to j.

$$DM^2 = \begin{bmatrix} dm_{aa} & dm_{ab} \\ dm_{ba} & dm_{bb} \end{bmatrix} \begin{bmatrix} dm_{aa} & dm_{ab} \\ dm_{ba} & dm_{bb} \end{bmatrix}$$

$$= \begin{bmatrix} dm_{aa}dm_{aa} + dm_{ab}dm_{ba} & dm_{aa}dm_{ab} + dm_{ab}dm_{bb} \\ dm_{ba}dm_{aa} + dm_{bb}dm_{ba} & dm_{ba}dm_{ab} + dm_{bb}dm_{bb} \end{bmatrix}$$

$$\tag{2.22}$$

$dm_{aa}dm_{aa} + dm_{ab}dm_{ba}$ is the sum of probability of staying at point a and probability of moving from point a to point b and then point b to point a. Increasing the powers of diffusion matrix increases the path length. DM_{ab}^k is the sum of all paths of length k from point a to point b [10].

Diffusion Process is the process of obtaining global connectivity between data points by combining the local connectivity which can be done by increasing the power of diffusion matrix. As the power increases, connectivity between more numbers of points can be obtained.

Diffusion distance is used to measure the similarity between two points in terms of their connectivity.

$$D_k(x_a, x_b)^2 = \sum_{u \in X} |p_k(x_a, u) - p_k(x_b, u)|^2$$

$$= \sum_l |DM_{al}^k - DM_{lb}^k|^2$$

(2.23)

where, u is any point in the dataset, $p_k(x_a, u)$ is the sum of probabilities of all possible k length paths between x_a and u.

Diffusion map tries to preserve the geometry of data by organizing data on the basis of diffusion matrix. For mapping data space coordinates to diffusion space coordinates diffusion distance simply becomes the Euclidean distance.

For

$$y_a = \begin{bmatrix} p_k(x_a, x_1) \\ p_k(x_a, x_2) \\ . \\ . \\ p_k(x_a, x_n) \end{bmatrix}$$

(2.24)

the Euclidean distance between two points y_a and y_b after mapping is

$$\|y_a - y_b\|_{Euclidean}^2 = \sum_{u \in X} |p_k(x_a, u) - p_k(x_b, u)|^2$$

$$= \sum_l |DM_{al}^k - DM_{lb}^k|^2 = D_k(x_a, x_b)^2$$

(2.25)

So Euclidean distance between points y_a and y_b is the diffusion distance between points x_a and x_b [10].

b) Normalization of the diffusion matrix

Suppose KM is a symmetric kernel matrix of size nxn. D is a diagonal matrix that contains row sums of KM. Matrix D normalizes the rows of kernel matrix KM to generate the diffusion matrix by multiplying each element of the matrix KM by the inverse of diagonal matrix D. Mathematically,

$$DM = KMD = D^{-1}KM \tag{2.26}$$

c) Compute eigenvectors and eigenvalues of diffusion matrix

Equation (2.24) can be represented with the help of eigenvectors and eigenvalues of normalized diffusion matrix as:

$$Y'_a = \begin{bmatrix} \lambda_1^k \psi_1(i) \\ \lambda_2^k \psi_2(i) \\ . \\ . \\ . \\ \lambda_n^k \psi_n(i) \end{bmatrix} \tag{2.27}$$

where, $\psi_1(i)$ is the i-th element of first eigenvector of DM and λ_1^k is the eigenvalue associated with the first eigenvector for path length k.

d) Mapping of d dominant eigenvectors and eigenvalues to the low dimensional space

Mapping to low dimensional space is done by selecting those eigenvectors that best reduces the difference between the Euclidean distance and diffusion distance. To obtain a d dimensional representation of d dominant eigenvectors are selected and the dimensions associated with these eigenvectors are preserved [10].

2.2.4 t-Distributed Stochastic Neighbor Embedding

Different dimension reduction methods try to preserve different properties of original data in lower dimension. Some techniques preserve the relationships between the data to keep local structure of data unchanged. Some other techniques try to keep similar data points close to each other and dissimilar data points far apart in lower dimension. Stochastic Neighbor Embedding (SNE) considers the neighboring points to obtain the probability distribution and try to approximate this value in lower dimension. t-SNE (t-Distributed Stochastic Neighbor Embedding) is a variation of SNE (Stochastic Neighbor Embedding) which helps in visualization of high dimensional data in low dimension. t-SNE captures both global and local structural information of the data. SNE finds the similarity between data points in the high dimensional space with the help of conditional probability which is calculated from the Euclidean distance [11]. The conditional probability $p_{b|a}$ represents the similarity between point x_b and point x_a. If the neighbors are selected on the basis of their probability density under a Gaussian centered at x_a, then $p_{b|a}$ represents the probability that x_a will select x_b as neighbor. Considering σ_a as the variance of the Gaussian function centered at x_a, $p_{b|a}$ can be represented as:

$$p_{b|a} = \exp(-||x_a - x_b||^2/2\sigma_a^2)\sum_{k \neq a}\exp(-||x_a - x_k||^2/2\sigma_a^2) \tag{2.28}$$

Suppose the low dimensional embedding of data points x_a and x_b are y_a and y_b respectively. Assuming variance in low dimensional space as $1\sqrt{2}$, the conditional probability in the low dimension becomes:

$$q_{b|a} = \exp(-||y_a - y_b||^2)\sum_{k \neq a}\exp(-||y_a - y_k||^2) \tag{2.29}$$

To obtain the low dimensional embedding of the high dimensional data SNE tries to minimize the difference between $p_{b|a}$ and $q_{b|a}$. Using gradient descent technique SNE tries to minimize the sum of Kullback-Leibler divergence over all data points.

For SNE the cost function C becomes:

$$C = \sum_a KL(P_a||Q_a) = \sum_a\sum_b p_{b|a}\log\frac{p_{b|a}}{q_{b|a}} \tag{2.30}$$

where, P_a is the conditional probability distribution over all points in the high dimensional space. Q_a is the conditional probability distribution over all points in the low dimensional embedded space.

The main issues in SNE that can be solved using t-SNE are the use of a cost function which is difficult to optimize and the crowding problem. The main differences between SNE and t-SNE are

– Unlike SNE, t-SNE uses symmetrized version of SNE cost function.
– In low dimensional space similarity between two points is calculated using student t distribution in t-SNE instead of Gaussian distribution of SNE [11].

t-SNE helps in solving both the optimization and crowding problem of SNE. t-SNE uses the concept of symmetric SNE by assuming $p_{ab} = p_{ba}$ and $q_{ab} = q_{ba}$, $\forall a, b$. Like SNE it also considers pairwise similarity. So, p_{aa} and q_{aa} both are made zero. Instead of considering the conditional probabilities, joint probability distributions can also be considered to minimize Kullback-Leibler divergence. Considering P as the joint probability distribution in high dimension and Q in low dimension equation (2.30) can be rewritten as:

$$C = KL(P||Q) = \sum_a\sum_b p_{ab}\log\frac{p_{ab}}{q_{ab}} \tag{2.31}$$

In symmetric SNE the similarity between two points in high dimensional space is:

$$p_{ab} = \exp(-||x_a - x_b||^2/2\sigma^2)\sum_{k \neq l}\exp(-||x_k - x_l||^2/2\sigma^2)$$

and in low dimensional space is:

$$q_{ab} = \exp(-||y_a - y_b||^2)\sum_{k \neq l}\exp(-||y_k - y_l||^2)$$

When a point x_a is an outlier in the high dimensional space, the pairwise distance with point x_b becomes large. This makes p_{ab} small for all b, as a result y_a has very little effect on cost function. This makes finding the position of a point from other points in low dimensional space difficult. To solve this problem joint probabilities are considered as symmetrized conditional probabilities, that means $p_{ab} = p_{b|a} + p_{a|b}2n$. This makes $\sum_b p_{ab} > 12n$ for all x_a, so that each point x_a plays an important role in finding the cost function [11].

The gradient of equation (2.30), that is for SNE is:

$$\delta C\delta y_a = 2\sum_b (p_{b|a} - q_{b|a} + p_{a|b} - q_{a|b})(y_a - y_b) \tag{2.32}$$

The gradient for symmetric SNE is simpler and faster than the asymmetric SNE and can be written as:

$$\delta C\delta y_a = 4\sum_b (p_{ab} - q_{ab})(y_a - y_b) \tag{2.33}$$

To reduce the effect of crowding problem probability distribution having much heavier tails than Gaussian distribution can be used in the low dimensional space to generate probabilities form distances. t-SNE uses a Student t-distribution with one degree of freedom in the low dimensional space as the heavy-tailed distribution. So, in low dimension joint probabilities can be defined as:

$$q_{ab} = (1 + ||y_a - y_b||^2)^{-1}\sum_{k \neq l}(1 + ||y_k - y_l||^2)^{-1} \tag{2.34}$$

Student t-distribution is nothing but an infinite mixture of Gaussians. Since there is no exponentiation operation, finding density of a point under Student t-distribution is faster than a Gaussian distribution.

Now, the gradient of the Kullback-Leibler divergence between P and Q (Student-t based joint probability distribution Q) is:

$$\delta C\delta y_a = 4\sum_b (p_{ab} - q_{ab})(y_a - y_b)(1 + ||y_a - y_b||^2)^{-1} \tag{2.35}$$

The experimental observations also prove that one advantage of t-SNE over SNE is that the gradient in t-SNE strongly repels data points which are dissimilar having small pairwise distances in low dimensional space. t-SNE also helps in detection of local optima [11].

2.2.5 Generalized Discriminant Analysis

Linear Discriminant Analysis does not work well if the problem is non-linear in nature. To solve this problem Generalized Discriminant Analysis is used which maps the input data to a high dimensional space of features having linear properties on which linear techniques like LDA can be applied. Suppose the input set X has N elements with C number of classes. x is a vector of input set and x^t is its transpose [12]. X_i is a subset of X containing n_i elements. So, $X = \bigcup_{i=1}^{C} X_i$ and $\sum_{i=1}^{C} n_i = N$. The covariance matrix,

$$INCOV = 1N \sum_{j=1}^{N} x_j x_j^t \tag{2.36}$$

Suppose the mapping between the input space X and the Hilbert space F is a non-linear mapping ϕ, so $x \rightarrow \phi(x)$. In the feature space the covariance matrix is

$$FEACOV = 1N \sum_{j=1}^{N} \phi(x_j)\phi^t(x_j) \tag{2.37}$$

Assuming the mean value of class i as $\bar{\phi}_i$, that means

$$\bar{\phi}_i = 1n_i \sum_{k=1}^{n_i} \phi(x_{ik}) \tag{2.38}$$

where, x_{ik} represents the element k of class i, the inter-classes inertia in feature space F can be represented as:

$$I = 1N \sum_{i=1}^{C} n_i \bar{\phi}_i \phi_i^t \tag{2.39}$$

So the covariance matrix becomes

$$FEACOV = 1N \sum_{i=1}^{C} \sum_{k=1}^{n_i} \phi(x_{ik})\phi^t(x_{ik}) \tag{2.40}$$

Where, *FEACOV* indicates the total inertia of the data into F.

The kernel function $k(x_a, x_b) = k_{ab} = \phi^t(x_a)\phi(x_b)$ on the Hilbert space F is considered to generalize Linear Discriminant Analysis. For classes p and q, the kernel function becomes $(k_{ab})_{pq} = \phi^t(x_{pa})\phi(x_{qb})$ (2.41)
Suppose K_{pq} is a $n_p \times n_q$ matrix and K is a $N \times N$ symmetric matrix such that $K_{pq}^t = K_{pq}$. Mathematically, $K = (K_{pq})_{\substack{p=1,....C \\ q=1,....C}}$, where, $K_{pq} = (K_{ab})_{\substack{a=1,....n_p \\ b=1,....n_q}}$ (2.42)

A block diagonal matrix W of size $N \times N$ is
$$W = (W_i)_{i=1,....,C} \tag{2.43}$$
Where, W_i is $n_i \times n_i$ matrix containing terms $1n_i$ [12].

Like LDA, GDA also tries to maximize inter-classes inertia and minimize intra-classes inertia. Maximization can be done by finding the values of eigenvalues$\lambda$and eigenvectors$v$from the equation

$$\lambda FEACOVv = Iv \qquad (2.44)$$

The largest eigenvalue of this equation gives the maximum value of the quotient between inter-classes and intra-classes inertia.

$$\lambda = v^t Iv v^t FEACOVv \qquad (2.45)$$

The eigenvectors can be represented as the linear combinations of feature elements as:

$$v = \sum_{p=1}^{C} \sum_{q=1}^{n_p} \alpha_{pq} \phi(x_{pq}) \qquad (2.46)$$

where,$\alpha_{pq}$are the coefficients and $v$represents the solutions which lie in the mapped feature space.
Equation (2.45) can also be rewritten as:

$$\lambda = \alpha^t KWK\alpha\alpha^t KK\alpha \qquad (2.47)$$

Suppose the eigenvector decomposition of matrix K is:

$$K = E\Gamma E^t \qquad (2.48)$$

where,$\Gamma$is a diagonal matrix containing nonzero eigenvalues, $E$is the matrix containing normalized eigenvectors associated with $\Gamma$and $E^t E = I$, for $I$is an identity matrix[12].
Using equation (2.48) in equation (2.47)

$$\lambda = (\Gamma E^t \alpha)^t E^t WE(\Gamma E^t \alpha)(\Gamma E^t \alpha)^t E^t E(\Gamma E^t \alpha) \qquad (2.49)$$

Let$\beta = \Gamma E^t \alpha$, equation (2.49) becomes

$$\lambda = (\beta)^t E^t WE(\beta)(\beta)^t E^t E(\beta) \qquad (2.50)$$
$$\Rightarrow \lambda E^t E\beta = E^t WE\beta$$
$$\Rightarrow \lambda\beta = E^t WE\beta \quad (\text{since}E\text{is orthonormal}) \qquad (2.51)$$

The projection of a point $z$can be computed from the normalized vectors $v$as

$$v^t \phi(z) = \sum_{p=1}^{C} \sum_{q=1}^{n_p} \alpha_{pq} k(x_{pq}, z) \qquad (2.52)$$

2.3 Related Work

Researchers have done a lot of work on the use of dimensionality reduction techniques. K. Yildiz et al. have performed a comparative study regarding the use of dimensionality reduction techniques on high dimensional data [13]. A. Sellami and M. Farah performed a comparative analysis on the use of dimensionality reduction techniques to interpret remote sensing images [14]. J. Song et al. applied structured sparse Principal Component Analysis to remove redundant features preserving the feature structure. This plays a very important role in object recognition by reducing the dimension of query image [15]. J. Wenskovitch et al. provided a detailed analysis on combining dimension reduction and clustering techniques for visual analysis [16]. Dimensionality reduction techniques also help in forecasting daily return of the stock market [17].

A. Durou et al. worked on measuring and optimizing the performance of an off-line writer identification system with the help of non-linear dimensionality reduction algorithms like KPCA (Kernel Principal Component Analysis), LLE (Locally Linear Embedding), Hessian LLE, Isomap, Laplacian Eigenmaps. Experiments were conducted on English handwriting (IAM dataset) and Arabic handwriting (ICFHR 2012 dataset) and it was observed that KPCA (Kernel Principal Component Analysis) gave better performance than others [18].

Kernel Principal Component Analysis does not consider the available fault information and only considers normal data for detecting faults. X. Deng et al. [19] proposed a technique of fault detection using KPCA (Kernel Principal Component Analysis) and KLNPDA (Kernel Local-Nonlocal Preserving Discriminant Analysis) models. Their proposed FDKPCA (Fault Discriminant enhanced KPCA) model monitors fault discriminant components along with non-linear kernel principal components for fault detection and to obtain better result.

KPCA technique was also used to extract non-linear relationship between multiple characteristics to solve MCPD (Multi-Characteristic Parameter Design) problems [20].

D. Kong et al. [21] applied k-PCA to fuse the sensitive features which are selected by the use of correlation coefficient method. The use of k-PCA helps in improving the speed during training and accuracy of prediction. The correlation between the actual tool wear and fused features is considered and the tool wear prediction model based on v-Support Vector Regression (v-SVR) is designed.

K-PCA was also used to reduce dimensions of features on which particle swarm optimized k-nearest neighbor technique was applied to detect bearing fault level [22].

Face features are extracted using k-PCA along with Support Vector Machine as the recognizer for face recognition [23]. J.-M. Lee et al. [24] applied k-PCA for process monitoring and found that the performance of k-PCA is better than linear PCA.

C. A. Thimmisetty et al. [25] applied diffusion maps to extract low dimensional information embedded in high dimensional space. Their proposed method combines manifold learning technique with Gaussian process regression to obtain non-linear correlations between the data points for sharpening high dimensional interpolation. The technique helps in predicting the geological properties. Diffusion mapping method was also used for dimension reduction along with the mahalanobis distance as a measure of similarity which plays very important role in fault detection [26]. Diffusion maps were used to solve the problem of defining differentiation trajectories [27]. Z. Farbman et al. [28] applied diffusion maps for different edge-aware operations which need to find the similarity between pixel pairs.

N. Patwari and A. O. Hero applied LLE, Hessian LLE and Isomap algorithms for localization in wireless sensor networks [29]. Multidimensional patterns can be easily understood by the use of manifold learning techniques, but these methods are not effective for sequential data and basically works in batch mode. H. Li et al. proposed an incremental method which can handle new samples each time. They proposed IHLLE (Incremental HLLE) which is an incremental version of HLLE [6].

When the dimension of feature space increases data points become more sparse. The sparse nature of data creates problem in proper extraction of hidden information. J. Birjandtalab et al. used t-Distributed Stochastic Neighbor Embedding (t-SNE) to embed high dimensional features in a low dimensional space [30]. A Gisbrecht et al. [31] proposed a variation of t-SNE called kernel t-SNE. Spatially mapped t-SNE was used to identify tumor subpopulations that can affect the survival of a patient in gastric cancer and metastasis status in the primary stage of breast cancer [32].

The performance of kernel-based learning methods depends on the kernel type and popularly used in classification, regression and dimension reduction. Z. Liang and Y. Li [33] proposed the use of multiple kernels for GDA (Generalized Discriminant Analysis). Generalized Discriminant Analysis was also applied considering the dissimilarity between multivariate observations [34].

Dimensionality reduction plays a very important role in image processing. Availability of huge storage devices at low price has increased the use of large image databases. Image processing is widely used in medical science, geographical information system, crime investigation, satellite and remote sensing image analysis, Engineering design and many more. For easy analysis, retrieval, classification and other image processing tasks features are extracted from the images. PCA, LDA are some of the techniques used for dimensionality reduction of images [35, 36, 37]. A. Sellami and M. Farah discussed on different feature extraction methods like PCA, TLPP (Tensor Local Preserving Projection), Kernel PCA, Laplacian Eigenmaps. A comparative study of dimensionality reduction methods was also performed considering remote sensing images. The comparative analysis considers band selection methods along with the feature extraction methods [38].

Tab. 2.1: Comparative analysis of dimension reduction techniques

Techniques	k-PCA	HLLE	Diffusion Maps	t-SNE	GDA
Parameters	Kernel function k (−, −)	Neighborhood size k	Connectivity, Path length	Variance σ	Covariance, Kernel function
Properties	– Capture structure of the data – Suitable for unsupervised, non-linear problems [12]	– Assumes a connected parameter space – Is a modification of LLE [6] – Theoretical framework has similarity with Laplacian Eigenmap, Laplacian operator is replaced by Hessian [7] – Gives better result than LLE – Not suitable for large manifolds due to high computational complexity	– Approximates the Euclidean distance in the embedded space with the diffusion distance in the feature space [10] – Kernel matrix is symmetric and positivity preserving	– Use symmetrized version of SNE cost function – Student t distribution is used – Optimization of t-SNE cost function is easier than that of SNE – No need to use simulated annealing like SNE to obtain good local optima [11]	– Finds optimal discriminant coordinates – Suitable for supervised, non-linear problems. [12]

Dimensionality reduction techniques are used in text mining to extract interesting and important information from unstructured text data. Singular Value Decomposition (SVD) [39] is used to reduce the dimension of textual documents. TF-IDF (Term Frequency-Inverse Document Frequency) [40] is a statistical measure to find important words from a document. WET (Weight of Evidence for Text), CHI Square are some other feature selection methods [40, 41]. Expected Cross Entropy (ECE), Information Gain (IG) are also used to select interesting features [41].

2.4 Experimental Observations

The surveyed non-linear techniques are tested on Diabetic Retinopathy Debrecen Dataset and Ecoli Data as shown in Figure 2.3 and Figure 2.4 respectively using the dimensionality reduction toolbox of MATLAB. The experimental results obtained for the five non-linear dimensionality reduction techniques using Diabetic Retinopathy Debrecen Data Set are given in Figure 2.5 and Ecoli Data in Figure 2.6. It is observed that t-SNE gives the best visualization as compared to others in Diabetic Retinopathy Debrecen Dataset. Data points are clearly distinguishable without any overlapping in the low dimensional space in case of t-SNE. The results are also analyzed using Ecoli Data as shown in Figure 2.6 and it is observed that on Ecoli Data k-PCA, Diffusion maps and t-SNE give clear visualization result. Comparative analysis of all the surveyed nonlinear dimensionality reduction techniques are shown in Table 2.1.

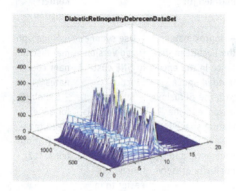

Fig. 2.3: Diabetic Retinopathy Debrecen dataset.

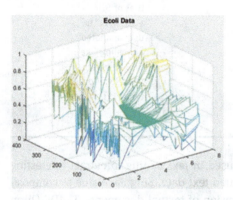

Fig. 2.4: Ecoli data.

2.5 Conclusion

Dimensional reduction is very important for easy analysis, better visualization of information present in an image or signal or real world data. The techniques for dimension reduction can also be combined with other techniques for getting better results. This chapter focuses on some of the nonlinear dimension reduction techniques along with the research work done using these techniques. A comparative analysis of these methods of dimension reduction has also been performed. The uses of dimension reduction in image processing and text mining are also discussed. The accuracy of any dimensionality reduction technique depends on the nature of the data. This can also be observed from the results obtained by applying k-PCA, HLLE, Diffusion Maps, t-SNE and GDA on Diabetic Retinopathy Debrecen Dataset and Ecoli Data. So, it can be concluded that dimension reduction techniques should be selected depending

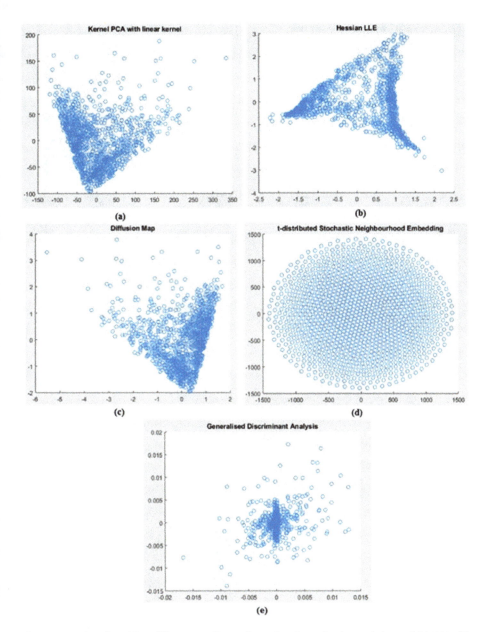

Fig. 2.5: Results of applying different non-linear dimensionality reduction techniques (a) k-PCA, (b) HLLE, (c) Diffusion Map (d) t-SNE (e) GDA on Diabetic Retinopathy Debrecen dataset.

on the type of data and reason of reduction. The surveyed techniques can also be applied on images and texts to verify their effectiveness and efficiency in reducing dimensions.

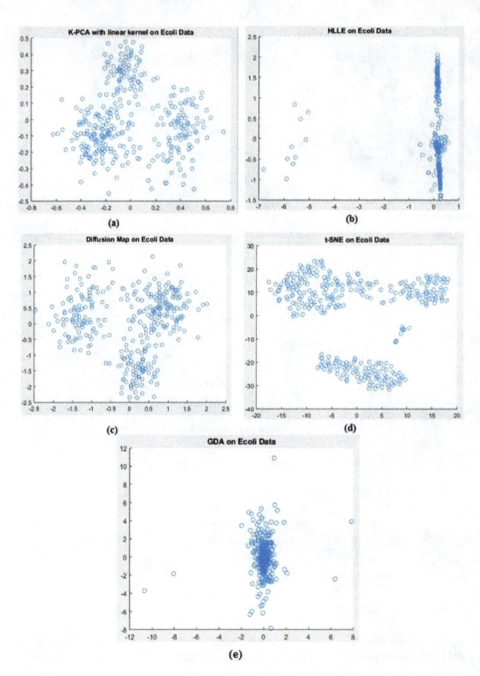

Fig. 2.6: Results of applying different non-linear dimensionality reduction techniques (a) k-PCA, (b) HLLE, (c) Diffusion Map (d) t-SNE (e) GDA on Ecoli data.

References

[1] Buchala, S., Davey, N., Gale, T. M. and Frank, R. J. 2005. Analysis of linear and nonlinear dimensionality reduction methods for gender classification of face images. International Journal of Systems Science, 36(14), 931–942.

[2] Gao, L., Song, J., Liu, X., Shao, J., Liu, J. and Shao, J. 2017. Learning in high-dimensional multimedia data: the state of the art. Multimedia Systems, 23(3), 303–313.

[3] VanDerMaaten, L., Postma, E. and VandenHerik, J. 2009. Dimensionality reduction: a comparative review.

[4] Holmström, L. 2008. Nonlinear dimensionality reduction by john a. lee, michel verleysen. International Statistical Review, 76(2), 308–309.

[5] Bishop, C. M. 2006. Pattern recognition and machine learning,Springer.

[6] Li, H., Jiang, H., Barrio, R., Liao, X., Cheng, L. and Su, F. 2011. Incremental manifold learning by spectral embedding methods. Pattern Recognition Letters, 32(10), 1447–1455.

[7] Xing, X., Du, S. and Wang, K. 2016. Robust Hessian locally linear embedding techniques for high-dimensional data. Algorithms, 9(2).

[8] Donoho, D. L. and Grimes, C. 2003. Hessian eigenmaps: Locally linear embedding techniques for high-dimensional data. Proceedings of the National Academy of Sciences, 100(10), 5591–5596.

[9] Coifman, R. R. and Lafon, S. 2006. Diffusion maps. Applied and computational harmonic analysis, 21(1), 5–30.

[10] DelaPorte, J., Herbst, B. M., Hereman, W. and VanDerWalt, S. J. 2008. An introduction to diffusion maps. In Proceedings of the 19th Symposium of the Pattern Recognition Association of South Africa (PRASA 2008), Cape Town, South Africa,15–25.

[11] Maaten, L. V. D. and Hinton, G. 2008. Visualizing data using t-SNE. Journal of machine learning research, 9(Nov), 2579–2605.

[12] Baudat, G. and Anouar, F. 2000. Generalized discriminant analysis using a kernel approach. Neural computation, 12(10), 2385–2404.

[13] Yildiz, K., Camurcu, Y. and Dogan, B. 2018. Comparison of Dimension Reduction Techniques on High Dimensional Datasets. International Arab Journal of Information Technology, 15(2), 256–262.

[14] Sellami, A. and Farah, M.2018. Comparative study of dimensionality reduction methods for remote sensing images interpretation. In 2018 4th International Conference on Advanced Technologies for Signal and Image Processing (ATSIP), 1–6. IEEE.

[15] Song, J., Yoon, G., Cho, H. and Yoon, S. M. 2018. Structure preserving dimensionality reduction for visual object recognition. Multimedia Tools and Applications, 77(18), 23529–23545.

[16] Wenskovitch, J., Crandell, I., Ramakrishnan, N., House, L. and North, C. 2018. Towards a systematic combination of dimension reduction and clustering in visual analytics. IEEE transactions on visualization and computer graphics, 24(1), 131–141.

[17] Zhong, X. and Enke, D. 2017. Forecasting daily stock market return using dimensionality reduction. Expert Systems with Applications, 67, 126–139.

[18] Durou, A., Aref, I., Elbendak, M., Al-Maadeed, S. and Bouridane, A. 2017. Measuring and optimising performance of an offline text writer identification system in terms of dimensionality reduction techniques. In 2017 Seventh International Conference on Emerging Security Technologies (EST), 19–25. IEEE.

[19] Deng, X., Tian, X., Chen, S. and Harris, C. J. 2017. Fault discriminant enhanced kernel principal component analysis incorporating prior fault information for monitoring nonlinear processes. Chemometrics and Intelligent Laboratory Systems, 162, 21–34.

[20] Soh, W., Kim, H. and Yum, B. J. 2018. Application of kernel principal component analysis to multi-characteristic parameter design problems. Annals of Operations Research, 263(1–2), 69–91.

[21] Kong, D., Chen, Y., Li, N. and Tan, S. 2017. Tool wear monitoring based on kernel principal component analysis and v-support vector regression. The International Journal of Advanced Manufacturing Technology, 89(1–4), 175–190.

[22] Dong, S., Luo, T., Zhong, L., Chen, L. and Xu, X. 2017. Fault diagnosis of bearing based on the kernel principal component analysis and optimized k-nearest neighbour model. Journal of Low Frequency Noise, Vibration and Active Control, 36(4), 354–365.

[23] Kim, K. I., Jung, K. and Kim, H. J. 2002. Face recognition using kernel principal component analysis. IEEE signal processing letters, 9(2), 40–42.

[24] Lee, J. M., Yoo, C., Choi, S. W., Vanrolleghem, P. A. and Lee, I. B. 2004. Nonlinear process monitoring using kernel principal component analysis. Chemical engineering science, 59(1), 223–234.

[25] Thimmisetty, C. A., Ghanem, R. G., White, J. A. and Chen, X. 2018. High-dimensional intrinsic interpolation using Gaussian process regression and diffusion maps. Mathematical Geosciences, 50(1), 77–96.

[26] Ma, W., Xu, J. and Li, Y. 2018. A Fault Detection Method with Mahalanobis Metric Learning Based on Diffusion Maps. In 2018 37th Chinese Control Conference (CCC), 5979–5984. IEEE.

[27] Haghverdi, L., Buettner, F. and Theis, F. J. 2015. Diffusion maps for high-dimensional single-cell analysis of differentiation data. Bioinformatics, 31(18), 2989–2998.

[28] Farbman, Z., Fattal, R. and Lischinski, D. 2010. Diffusion maps for edge-aware image editing. In ACM Transactions on Graphics (TOG), 29(6).

[29] Patwari, N. and Hero, A. O. 2004. Manifold learning algorithms for localization in wireless sensor networks. In 2004 IEEE International Conference on Acoustics, Speech, and Signal Processing, (ICASSP'04). 3. IEEE.

[30] Birjandtalab, J., Pouyan, M. B., Cogan, D., Nourani, M. and Harvey, J. 2017. Automated seizure detection using limited-channel EEG and non-linear dimension reduction. Computers in biology and medicine, 82, 49–58.

[31] Gisbrecht, A., Schulz, A. and Hammer, B. 2015. Parametric nonlinear dimensionality reduction using kernel t-SNE. Neurocomputing, 147, 71–82.

[32] Abdelmoula, W. M., Balluff, B., Englert, S., Dijkstra, J., Reinders, M. J. T., Walch, A., McDonnellL. A. and Lelieveldt, B. P. F. 2016. Data-driven identification of prognostic tumor subpopulations using spatially mapped t-SNE of mass spectrometry imaging data. Proceedings of the National Academy of Sciences, 113(43), 12244–12249.

[33] Liang, Z. and Li, Y. 2010. Multiple kernels for generalised discriminant analysis. IET Computer Vision, 4(2), 117–128.

[34] Anderson, M. J. and Robinson, J. 2003. Generalized discriminant analysis based on distances. Australian & New Zealand Journal of Statistics, 45(3), 301–318.

[35] Shereena, V. B. and Julie, M. D. 2015. Significance of dimensionality reduction in image processing. Signal & Image Processing, An International journal (SIPIJ), 6(3), 27–42.

[36] Kavzoglu, T., Tonbul, H., Erdemir, M. Y. and Colkesen, I. 2018. Dimensionality reduction and classification of hyperspectral images using object-based image analysis. Journal of the Indian Society of Remote Sensing, 46(8), 1297–1306.

[37] Bhujle, H., Vadavadagi, B. H. and Galaveen, S. 2018. Efficient non-local means denoising for image sequences with dimensionality reduction. Multimedia Tools and Applications, 1–19.

[38] Sellami, A. and Farah, M. 2018. Comparative study of dimensionality reduction methods for remote sensing images interpretation. In 2018 4th International Conference on Advanced Technologies for Signal and Image Processing (ATSIP), 1–6. IEEE.

[39] Kumar, A. A. and Chandrasekhar, S. 2012. Text data pre-processing and dimensionality reduction techniques for document clustering. International Journal of Engineering Research & Technology (IJERT). 1.

[40] Akkarapatty, N., Muralidharan, A., Raj, N. S. and Vinod, P. 2017. Dimensionality reduction techniques for text mining. In Collaborative Filtering Using Data Mining and Analysis. 49–72. IGI Global.

[41] Lu, Z., Yu, H., Fan, D. and Yuan, C. 2009. Spam Filtering Based on Improved CHI Feature Selection Method. In 2009 Chinese Conference on Pattern Recognition. CCPR 2009. 1–3. IEEE.

Makarand Velankar, Amod Deshpande, and Parag Kulkarni

3 Application of Machine Learning in Music Analytics

Abstract: With the growth of the internet and social media, music data is growing at an enormous rate. Music analytics has a wide canvas covering all aspects related to music. This chapter provides a glimpse of this large canvas with sample applications covered in detail. Machine learning has taken a central role in the progress of many domains including music analytics. This chapter will help the readers to understand various applications of machine learning in computational musicology. Music feature learning and musical pattern recognition give conceptual understanding and the challenges involved. Feature engineering algorithms for pitch detection or tempo estimation are covered in more detail with available popular feature extraction tools. Music classification and clustering examples explore the use of machine learning. Various applications ranging from the query by humming to music recommendations are provided for efficient music information retrieval. Future directions and challenges with deep learning as a new approach and incorporation of human cognition and perception as a challenge makes this domain a challenging research domain.

Keywords: Computational music, feature engineering, pattern recognition, music classification, and clustering, machine learning

3.1 Introduction

Music data analytics has gained substantial attention with the growth of music data across various modes such as the release of new albums, social media, the internet, etc. It involves online music streaming, music purchases, uploads, on-line discussions, and comments, etc. Automatic music recognition tasks such as instrument or artist, style, genre, emotions, melody, etc. have become necessary with the voluminous growth of music. The manual effort involves a significant time of musical experts and it is almost impossible for humans to do such tasks for massive growing musical data. Today, music lovers or creators generate and upload musical files over the Internet in audio or video form. This music data is growing at an enormous rate with the use of smartphones and different applications. Automatic computer-generated music with artificial intelligence is also contributing to the fast growth of music data. Approaches involving machine learning [Alpaydin, E., 2014], social network analysis [Borgatti et. al. 2018], signal processing [Smith 2011], etc. with feature engineering are being used for various applications in music analytics. Music analytics applications in music information retrieval involve copyright monitoring, music transcription, optimized search, play-list generation, personalized systems,

https://doi.org/10.1515/9783110610987-005

music recommendation, top charts, hit song prediction, etc. Current approaches rely mainly on music metadata and user inputs. These approaches are not scalable and are highly subjective. Content-based approaches or hybrid approaches are being developed to cater to the needs [Thorat et.al.2015].

Machine learning has taken a central role in various domains due to its ability to improve performance with experience [Michalski et al. 2013]. Features and classes are an integral part of machine learning. The choice of features is crucial in the performance of the tasks. A class is a group of similar elements. Models used for learning from labeled data are considered as supervised and unlabeled data as unsupervised learning models. Classification and clustering are typical tasks in various fields. The use of appropriate features and models is necessary for the success of a particular task. Pattern identification is a common task required in different applications and measuring pattern similarity is essential in clustering and classification of different patterns.

Patterns are dominant in music with representative examples as melodic or rhythmic patterns [Shmulevich et al.2001]. Music has a global presence and universal appeal, which makes it an important domain for various studies. The growth of music data in recent years has led to an increase in the need for applications to handle it effectively. Music data growth accelerated in recent years due to the use of social media and the internet for uploading and accessing musical data. Music data analytics is a widespread field with music streaming, promotion, analysis, recommendation, purchases including all activities related to music. Machine learning is used in various music analytics applications such as music information retrieval, recommendations, analysis, top-chart predictions, music generation, education, classification, transcription, etc. Music has multiple dimensions such as melody, rhythm, timbre, genre, instrumentation, mood, style, etc. Feature selection is a challenging task in music analytics due to the complex nature of music.

Covering the entire spectrum of machine learning applications for music analytics [Lerch, A., 2012] is beyond the scope here. Representative examples are provided to appreciate the use of machine learning in music analytics. This chapter focuses on music audio and related perceived features in machine learning applications for music analytics. It explores the need and role of machine learning in music analytics and covers some representative features to explore the use of machine learning in this ocean of music analytics. The chapter is organized in the following manner. Section 3.2 provides brief about music features, tools and feature extraction with sample examples for tempo estimation in detail. Musical pattern recognition is covered in section 3.3 with exploring melodic patterns and similarity estimation. Section 3.4 covers classification and clustering as typical machine learning applications for music analytics. Section 3.5 explores some of the applications in computational music. Future directions and challenges in computational music and deep learning as a new paradigm in machine learning are covered in section 3.6.

3.2 Features of Music Analytics

Computational music analysis involves feature engineering for music. The objective of music feature engineering is to mimic human perception and understanding music dynamics. It helps to generate and use relevant features for the specific task. Music is a rich harmonic audio signal [Datta et.al.2017] with a variety of forms and musical dimensions. The huge canvas of music evolved through centuries and decades involves a variety of music genres and features. Feature engineering is one of the crucial steps in any machine learning application [Thickstun, 2016]. Better the features, better the results, and accuracy in machine learning applications. Music data is stored in different forms which are termed as multimodal information. Music audio, lyrics, meta-data, notations, video along with articles, reviews, ratings, comments on music are components of music data analytics. Perceived musical features by humans involve musical interpretation and musical knowledge. Typical perceived musical features are melody, rhythm, artist, instrument, emotions or mood, etc. Melody or tune is characteristic of musical composition with sequence and duration of notes in a specific order. Notes are associated with pitch values measured in hertz. Rhythm is typically a repeated sound pattern associated with the tempo or speed of the music. The tempo is measured in BPM or beats per minute. Timbre is a source of the musical sound by which one can recognize a specific instrument or artist. Music is intended to produce specific emotions or create some mood in the mind of listeners. These emotions are subjective and are the outcome of various factors such as musical or cultural background likes and dislikes cognitive interpretation and psychological experience. Circumflex model is used by the majority of researchers to model different emotions on the valence arousal axis.

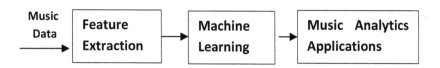

Fig. 3.1: Music Analytics System

Figure 3.1 provides a typical flow of the process in music analytics applications. Feature extraction involves extracting useful features from the music data. The music data used here are audio files. Other forms such as MIDI files, lyrics, notations, etc. are also useful in extracting appropriate features. The machine learning model is used to train the system for a specific application using the features extracted. This section provides the breadth with feature types, tools used and an example for tempo feature extraction. A representative example of tempo estimation is provided as a flavor in music feature engineering. This example will help to understand the complexity involved in feature engineering for the musical feature spectrum. Various feature generation techniques for music audio data are provided in documentation with a list of available tools and samples about features extracted from various popular tools.

3.2.1 Feature extraction tools

Different feature extraction tools are developed by MIR (Music Information Retrieval) community for various applications such as analysis, visualization, audio effects, etc. [smcnetwork]. A list of some of the music analysis tools and features supported are mentioned in Table 3.1 as a reference. Different features supported by these tools provide feature categories associated with music. These tools have already built-in tested algorithms to extract different features. Most of the tools are free with some exception such as MATLAB. Supporting examples and the instructions are provided in the documentation available for the tools.

Tab. 3.1: Music feature analysis tools

Sr.no.	Tool	Features supported
1	MIR Toolbox in MATLAB	Tonality, rhythm, structures
2	jMIR	Symbolic, Various timbral, spectral
3	Sonic Visualizer	Spectrogram, beat tracker, pitch tracking
4	Praat	Pitch, intensity, formant
5	Essentia	Tonal, rhythmic, time, spectral descriptors
6	Aubio	Silence, pitch, onset detection
7	Meloadia	melodic pitch contour

3.2.2 Feature extraction and learning

Feature learning attempts to make the machine learn the features to perform specific tasks. It can be broadly categorized into supervised and unsupervised. Supervised learning uses labeled data and unsupervised learning uses unlabeled data. Different tools mentioned in the previous subsection are used to extract features from the musical audio to train the machine to perform a specific task. Music features are extracted using the window of certain milliseconds typically 20 to 40 ms and ultimately aggregated over the entire audio file to represent global features. Depending on the application, local features extracted in a specific window may be useful for microlevel analysis.

Automatic timbre identification or instrument/voice identification is useful in applications like music information retrieval, automatic annotation of audio files, music segmentation, etc. Timbre is referred to as color or tone quality of sound by which one recognizes specific singers [Deshmukh et al.2015] or instrument [Bhalake et al. 2016]. Mel frequency spectral coefficients (MFCC) features, ADSR envelope (attack, decay, sustain, release) features are useful for timbre identification. Tempo estimation is another challenging task that involves different algorithms to estimate the tempo of the music. The tempo is measured in beats per minute (BPM) and is associated with the rhythm. To determine the tempo, various techniques such as autocorrelation and novelty curve for onset detection, beat tracking, etc. [Grosche & Muller 2011] are widely used. Research for real-time beat tracking has achieved pioneer results as well [Oliveira, et al 2011]. A performance of Raga Malkauns was analyzed by some of these algorithms.

Figure 3.2 and Figure 3.3 show the tempograms for Raga Malkauns. Tempogram is a representation of tempo with respect to time. These tempograms are computed using the Matlab Tempogram Toolbox [Grosche & Muller 2011]. No real tempo information can be extracted before 450 seconds as can be noticed from Figures 3.2 and 3.3. The reason for this irregularity is the absence of any rhythmic structure in Alap. A definitive tempo is predicted after the rhythm starts. Some irregularities are observed near 850, 1150 and 1300 seconds. An increase in tempo can be clearly observed after approximately 1300 seconds. These irregularities also correspond to an increase in energy from Figure 3.3. A combination of the information could lead to understanding of rhythm solos. Such solos are common in Hindustani Classical music.

A clearer picture can be observed in Figure 3.4. The variation in tempo is plotted against time. The INESC Porto Beat Tracker (IBT) algorithm [Oliveira, et al 2011] is used. Consistency in the results from these algorithms implies the algorithms can be used for Hindustani Classical music interchangeably depending on the use case. To compare the tempo variation in different music genres, three other songs were analyzed using the IBT algorithm. Song 1 is a progressive rock song having mild tempo changes and song 2 being mainstream popular Hindi song has even fewer tempo changes. Song 3 is a mainstream Indie rock song that has a structure similar to Hindustani Classical music but has a constant tempo throughout. The variation in tempo with time can be observed from Figures 3.5, 3.6 and 3.7 respectively. Unlike the Hindustani Classical composition, the tempo appears to be more constant in the first two scenarios. One would expect similar stability in Figure 3.4 as the tempo is constant in the stable regions. This can be attributed to the way the rhythm instruments such as Tabla are played. The conventional onset detection algorithms fail to give an accurate tempo in this case. The structure with a lack of rhythmic instruments makes it difficult to detect tempo which can be observed in Figure 3.7.

The median tempo analysis as shown in Table 3.2 gives an idea about the tempo of the composition as a whole. This method proves to be satisfactory for song 3, thus tackling the problem of lack of rhythmic instruments in the introduction. But it again proves to insufficient in the case where the tempo changes.

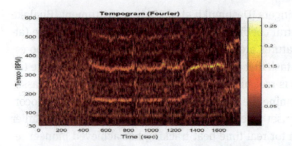

Fig. 3.2: Tempogram of Raga Malkauns, Fourier

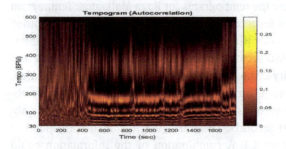

Fig. 3.3: Tempogram of Raga Malkauns, Autocorrelation

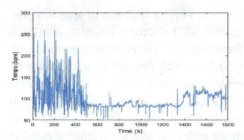

Fig. 3.4: Variation in Tempo in Raga Malkauns using IBT algorithm

Tab. 3.2: Median tempo of the compositions in BPM

Composition	Tempo (BPM)
Song 1	161
Song 2	96
Song 3	152
Raga Malkauns	92

Tempo feature extraction shown with examples provides complexity involved and challenges related to feature extraction in music. Features are critical for the success of any application in music analytics. The tempo associated with rhythm provides an idea about the rhythmic patterns in music. Music is occupied with patterns like repetitive structures with an emphasis on the specific pattern to convey specific musical meaning.

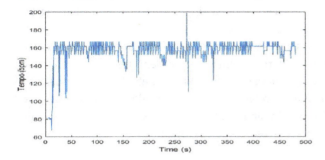

Fig. 3.5: Variation in tempo in song 1, IBT

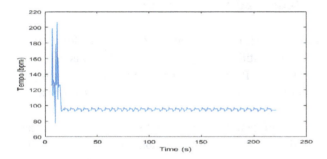

Fig. 3.6: Variation in tempo in Song 2, IBT

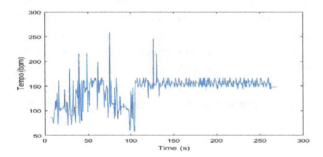

Fig. 3.7: Variation in tempo in song 3, IBT

3.3 Pattern recognition for Music

Patterns are everywhere and are assumed to be some repetitive tasks or predefined tasks in general. Pattern recognition is also associated with some co-relation or association existing in the data. Music is full of different patterns such as typical structural patterns of melody and rhythms. Music data mining may reveal unknown patterns or relationships present, which are not obvious to identify. Automatic recognition of musical patterns such as melodic or rhythmic patterns is a challenging task considering the complexity of musical data and feature extraction.

Pattern recognition in music is an important task for content-based retrieval to find similar patterns. Melodic pattern identification is useful for various tasks such as retrieval, recommendation, plagiarism issues in music, etc. Modeling music similarity based on melodic patterns involves challenges considering multidimensional musical features and subjective human perception. As pathways for overcoming these challenges, melodic pattern representation methods for melodies are discussed.

3.3.1 Algorithms for melodic pattern recognition

Melody is the heart of music and extracting melodic features is essential for a variety of applications such as similarity identification, copyright infringements, query by humming, content-based retrieval, etc. Pitch estimation is important in identifying melodic contour. Melodic contour symbolizes changes in perceived pitch over a definite time frame. Various pitch detection algorithms (PDA) based on time or frequency domains are available are used to identify melodic contour [Rao 2010]. Different factors considered and affects the PDA based on perception are fundamental pitch estimation, overtones, just noticeable difference (JND), silence, shimmer, jitter, etc. A typical melodic pattern extracted using feature extraction tool sonic visualizer is as shown in Figure 3.8. The x-axis represents the timeline (selection of 4.334 seconds) and Y-axis shows frequency in Hz (changes in the melodic pattern). Pitch values in frequency are converted to notations to represent the melody in notation form.

For the implementation, pitch extraction as mention in the previous section was performed. The pitch contour was extracted from 'Praat' in an 'Excel' sheet. Along with this, the Just Intonation filter bank was created using the tonic and its frequency value in the lowest octave. This was possible as pitch extraction was already performed.

3.3.2 Melodic pattern representation and similarity

Melodic patterns are represented in structural patterns as music notations or MIDI (Musical Instrument Digital Interface) format [Arifi et al. 2003]. Music notations are represented in different forms such as sheet music for western music or bhatkhande notations for Indian music. Different music traditions have their own forms of representation of melody in symbols or notations.

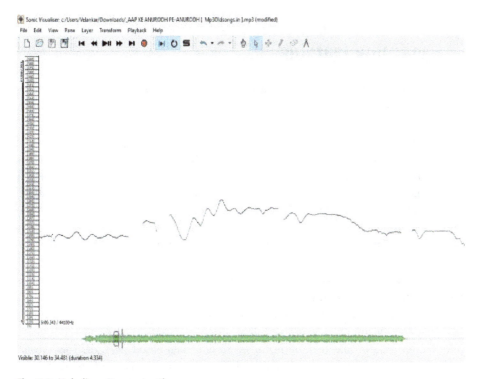

Fig. 3.8: Melodic pattern extraction

Representation of melody using western notation sequence is noted in a specific sequence in the melody. The example of the melody using notations for the song 'Jingle bell' is E EE, E EE, E G C D. The same can be represented using relative representation considering the transition and distance of notes in the octave is 0 0 0 0 0 0 +3 −7 +2. Here 0 represents no transition or same note repetition and +3 represents note transition with the change to 3 notes ahead in the octave whereas -7 represents a change to note at 7th position before the current note. Melodic pattern matching is like matching strings in linguistics.

Melodic pattern recognition for Indian raga music has been one of the most researched topics in Indian computational music. Challenges in raga pattern identification along with different approaches are explored and experimentation revealed the use of pitch class distribution (PCD) and n-gram approach as a computationally efficient and useful approach for pattern recognition [Velankar et al. 2018]. The PCD is a distribution of the probability of the occurrence of these pitches. An n-gram represents the sequence of the occurrence of 'n' notes. Thus, a 2-gram would represent a two-note sequence and a 3-gram would represent a three-note sequence. N-grams showcase important information about the melody structure. PCD and N-grams are used to train the machine learning model for melodic pattern recognition.

3.4 Music classification and clustering using machine learning

Various feature extraction tools along with feature learning [Bisot 2017] algorithms are used to train the model and test for application. Machine learning with supervised or unsupervised approach helps to learn specific phenomenon and perform the task efficiently. Machine learning has a wider role to play in automatic extraction of various features, feature learning using different machine learning approaches, pattern recognition, similarity identification, classification, and clustering. Classification and clustering are typical tasks used in machine learning along with prediction. Different learning approaches are being used to perform these tasks. The supervised learning approach uses labeled data to train the machine learning model. In the case of unsupervised learning, no labels are available/ provided for the data objects. Other learning approaches such as semi-supervised learning uses a combination of labeled and unlabeled data. Reinforcement learning uses a reward-punishment methodology to train the model.

Machine learning in music analytics involves training the machine generally using the supervised algorithms for classification purposes. Practical applications of music information retrieval are majorly restricted to metadata-based features. This creates both a void and an opportunity. Training machine to learn the content-based features is important from the semantic view for humans and is a challenging task. Efficient retrieval based on content-based features is essential for music lovers. Approaches for popular applications like genre, instrument and emotion classification [Aljanaki 2017] are being improved with the use of machine learning.

In supervised feature learning, these features are extracted for different instruments or voice samples to train the model with labeled data and further tested for the unknown samples to test the accuracy in training and testing of the model used. Different widely used supervised algorithms include naïve Bayes, support vector machine, decision trees, multilayer perceptron (neural networks), etc. A typical supervised approach is explained with the block diagram as shown in Figure 3.9. Labeled song database is used as a source, which is divided into training and test datasets with a proportion of 80:20 generally. 80 percent of samples are used for training and the remaining 20% are kept for testing the learned model. Feature extraction is done using available tools or developing different algorithms. Relevant features are selected to train the machine learning classifier. The model is tested using test datasets to verify the accuracy of the classifier. Unsupervised learning is generally applied for a large collection of music for the automatic clustering of musical data when data is not labeled. It helps in grouping similar music items together for efficient search and retrieval.

An example with a sample dataset of 50 songs is used to illustrate the use of learning of features for classification and clustering. These 50 songs are classified

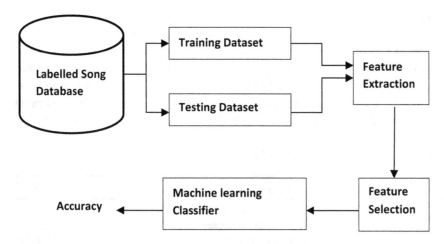

Fig. 3.9: Typical supervised learning approach for classification

into 5 classes as per manual labeling as shown in Table 3.3. The same dataset is used for the experimentation with machine learning.

Tab. 3.3: Dataset for classification with manual labeling

Class	Elements	Data objects as music audio files
1	15	1, 3, 10, 13, 17, 27, 28, 30, 31, 33, 36, 40, 41, 43, 45.
2	17	2, 5, 6, 7, 8, 9, 11, 12, 14, 16, 19, 21, 23, 24, 25, 29, 32.
3	4	4, 15, 18, 46.
4	3	20, 22, 37.
5	11	26, 34, 35, 38, 39, 42, 44, 47, 48, 49, 50.

3.4.1 Supervised approach for classification

Automatic music classification is the need considering the exponential growth of music data in recent times and future predictions. Various music classification tasks include classification based on genre, artist, style, instrument, mood [Hu 2017], etc. The dataset referred to in Table 3.3 is used for testing machine learning classifier accuracy. 72 features are extracted using jaudio toolbox for each audio file. The subset evaluator approach is used to evaluate the value of a subset of attributes by considering a specific analytical capability of each feature along with the degree of redundancy between them. Subgroups of features that are highly associated with the class while having a little intercorrelation are preferred. This feature selection approach provided 11 features from a total of 72 features. A support vector machine (SVM) is used as a supervised classifier to train and test the model using 10-fold cross-validation. A 10-fold cross-validation methodology is used to evaluate the accuracy of the classifier for a specific feature set. Different parameters are used for determining the accuracy of a classifier. Correctly a classified instance from the dataset is one of the simplest measures which provides accuracy in percentage. Classification accuracy is one of the important measures for the comparison of performance. Other measures include root mean square error, relative absolute error, precision, recall, F-measure, ROC area, etc. A class-wise accuracy can be observed using a confusion matrix. An example of the SVM classifier result is as shown in Table 3.4.

Tab. 3.4: Accuracy measures for SVM classifier

Sr. No.	Accuracy Measure	Value
1	Correctly classified instances	44
2	Accuracy in percentage	88%
3	Mean absolute error	0.2456
4	Root mean squared error	0.3248
5	Relative absolute error	82.26%
6	Root relative squared error	84.16%
7	Precision	0.893
8	Recall	0.880
9	F-measure	0.887
10	ROC Area	0.846

Another important visualization is a confusion matrix (error matrix) which helps to understand a difference between the actual classification and predicted classification for each class element which can further be used to find true positives, true negatives, false positives, and false negatives. Table 3.5 provides a confusion matrix for the SVM classifier. It shows incorrect predictions as 1, 0, 1, 2 and 2 for classes 1 to 5 respectively. Accuracy is calculated for a total of 6 improper predictions out of a total of 50 as 88% with 44 correct predictions. An ideal confusion matrix is a diagonal matrix with only diagonal elements for a principal diagonal and all other elements as 0.

Tab. 3.5: Confusion matrix for SVM classifier

Predicted Class Actual Class	1	2	3	4	5
1	14	1	0	0	0
2	0	17	0	0	0
3	0	0	3	1	0
4	0	2	0	1	0
5	0	2	0	0	9

3.4.2 Unsupervised approach for clustering

The unsupervised approach uses an unlabeled dataset and uses different clustering algorithms to classify the songs Different unsupervised learning algorithms are available to explore the unlabeled data to identify hidden patterns and relationships. K means clustering is one of the popular unsupervised learning algorithms which attempts to classify data into mutually exclusive k clusters. The value of k is experimentally decided or empirically decided considering the data items to be clustered. This is an iterative algorithm to assign each data item to a specific cluster on the basis of features selected. Table 3.6 shows the clustering of the same data referred to in Table 3.3 with no labels provided. Comparison of Tables 3.3 and 3.6 can reveal some interesting hidden patterns. As can be noticed some of the songs are classified into different groups. Further introspection of these songs revealed that they can be classified into both classes and there is a need for multi-class labeling for the songs instead of unique class labeling.

Tab. 3.6: Unsupervised clustering using k means clustering algorithm for k=5

Class	Elements	Data objects as music audio files
1	18	1, 2, 5, 6, 7, 8, 9, 11, 12, 14, 16, 19, 21, 23, 24, 25, 32, 38
2	4	20, 22, 37, 44
3	3	4, 15, 18
4	13	26, 28, 34, 35, 39, 40, 41, 42, 43, 45, 47, 48, 49
5	12	3, 10, 13, 17, 27, 29, 30, 31, 33, 36, 46, 50

Clustering is viewed in different machine learning tools using visualization techniques available. Figure 3.10. shows cluster visualization for 4 clusters using unsupervised learning with data points for each cluster shown in different colors using tool weka. It shows 50 data points scattered across the graph for a specific selection of features as 14, 30, 34, 69 and 70 for the clustering by removing other features. X and Y-axis denote specific features and different feature selection for axis will show different views. Visualization helps in understanding the spread and data points in each cluster formation.

3.5 Applications in computational music

Machine learning has been used in various applications in music analytics and has a wide research potential to use it furthermore effectively to build new enhanced systems and applications. Some of the applications are explained in the following section to give a glimpse.

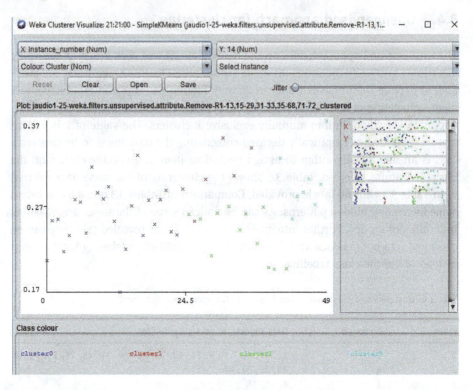

Fig. 3.10: Visualization of clustering using tool weka

3.5.1 Machine learning for music information retrieval

Music information retrieval [Liem et al. 2011] community focuses on the research and development of computational systems to help humans make better sense of music data. Different applications such as automatic genre classification, music emotion recognition, recommendation, playlist generation, a ranking of songs have been developed and tested with various suitable approaches such as multi-class, multi-label, semi-supervised, deep learning, etc. A typical system for music information retrieval is presented using Figure 3.11. Musical features extracted are used to train the machine learning model which further provides solutions for proper indexing and ranking mechanism. The objective is to provide appropriate relevant results for the user queries. The International Society for Music Information Retrieval [ismir] is working on the research and development of computational systems since the last decade to help humans better make sense of music data.

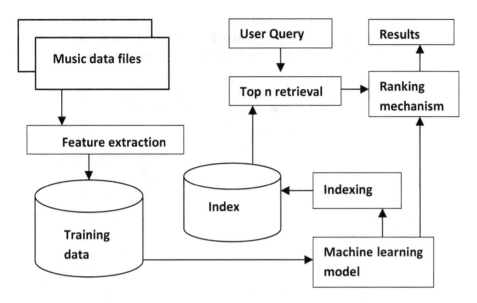

Fig. 3.11: Typical machine learning system for music retrieval

3.5.2 Query by humming

Query by humming is a natural way of querying the song and melodic information extraction and matching from the large music corpus [Serra 2014, Panteli et al. 2018] is a challenge. Searching a song on the internet with a huge database of songs uses a meta-data-based approach mainly with search based on text. The humming of the song is a possible natural way of expressing query and search a song from the musical database. Query by humming (QBH) system stores melodic information of the songs to match with humming pattern to retrieve the best possible match or matches with ranking [Makarand V 2018]. The typical query by humming system is as shown in Figure 3.12. The main drawback of the QBH system is the scalability issue and the need for manual annotation to build a database. With automatic transcription algorithms gaining acceptable results, manual annotation will not be required and automatic database generation is possible. As per findings, none of the automatic transcription algorithms is absolutely error-free and it limits the practical usage for QBH System. Melodic contour is stored by extracting features such as pitch information of a musical file [Salamon, J. and Gómez, E., 2012]. Other features such as duration, intensity associated with the melody are extracted and stored as a feature vector. The study of the query by untrained singers is useful for the decision of minimum duration for note perception. It is about 100 milliseconds as per the observations. Notes presented during query by humming are not likely to be flawless as singers and they are possible to differ from the original frequencies and durations. For such deficient queries to accommodate possible faults in the query, a greater range of frequencies keeping the actual frequency at the center to represent the particular note needs to be considered. In QBH application, this range should be more to detect the possible frequencies in user query considering non-singers.

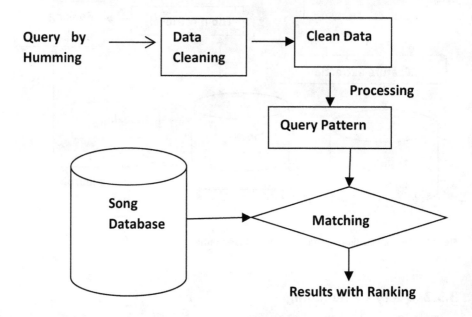

Fig. 3.12: Typical Query by humming system

The dynamic time wrapping method is used for time-stretched or compressed queries to accommodate duration errors in user queries. The note sequence for the tune 'Jingle bell' with piano notes starting with E (E EE, E EE, E G C D) is also represented by shifting one key and starting note F as (F FF, F FF, F G# C# D#). The user presents a query in any octave with a different starting note. The same melodic sequence is convenient to represent in the form of relative notes transition considering their positions on both sides. No change (playing the same note) marked as 0, the transition towards right and left side is considered as positive and negative steps respectively (steps represent the position of new note from current one). The relative representation for song 'Jingle bell' is represented as (0 0 0 0 0 0+3 −7 +2). Here transitions, 0 indicates the same key (E to E); +3 as 3 position towards right from the present key (E to G) and -7 as 7 positions towards left from the present key (G to C), etc. This relative note representation is feature engineering applied for the QBH application. Matching techniques such as edit distance or n-grams are used to find the best match or matches to rank the results. Domain expert knowledge helps in deciding suitable query matching algorithms considering different music acoustics aspects and perception of similarity. Matured QBH systems will be developed with the help of machine learning by training the system to adopt the errors in the user queries and provide the best possible results.

3.5.3 Music plagiarism and similarity

Plagiarism in music is a copy of the musical artwork which is generally the composition or tune. An example is generally a composer accusing the other of copying and/or reproducing their work for financial gains without consent. It is claimed as inspiration by the other composer and the line is very thin between copy and inspiration. It is difficult to conclude and prove in true sense. The system can, however, compute the similarity between 2 compositions with statistical support. Music similarity is a hard problem when considered as a general similarity. It is solvable to the extent if the similarity is considered for specific aspects such as tune or rhythm or artist or instrument used etc.

Court decision on music plagiarism and role of music similarity has been explored [Müllensiefen, D. and Pendzich, M., 2009] and it was proposed that predictive values of music similarity algorithms can be useful for the decision. [Dittmar et al.] proposed a toolbox with audio forensics for the inspection of suspected music plagiarism cases. This is a widely debated and controversial topic considering the money involved in the music industry. Technology advances will further assist in making appropriate decisions related to music plagiarism in the coming future.

3.5.4 Recommendation systems

A music recommender system is a system that learns from the user's past listening history and recommends them songs which they would probably like to hear in the future. The present system of recommendation is based on the Collaborative filtering and popularity of songs. The recommendation is based on the user's preference from historical usage and user profile matching. This recommendation system is based on the popularity of the songs i.e. likes and most visited songs influence the recommendation. But the biggest problem is new and unpopular songs that cannot be recommended due to lack of past data as it is difficult to predict the taste of the newly registered user. This is referred to as the cold start problem. Music recommendation systems today are mainly based on collaborative filtering which has issues like cold start problems. Another issue in the recommendation system is algorithmic bias. In a real-world music recommendation system, where due to a problem of measuring distances in high dimensional spaces, songs closer to the center of all data are recommended over and over again, while songs far from the center are not recommended at all.

A content-based recommendation is an approach likely to resolve issues in a collaborative filter approach. Different approaches are evaluated with the use of machine learning in recommendation systems, such as based on music sentiments [Rosa et al. 2015], location-aware retrieval [Schedl, M. and Schnitzer, D., 2014], hybrid approach [Wang, X. and Wang, Y., 2014], etc. These approaches attempt to improve the efficiency in music recommendation systems by adopting user-specific recommendations with the profiling of users and analyzing musical contents.

3.6 Future directions and Challenges

Deep learning has remarkable success in the area of speech and image recently and researchers are using deep learning for computational music for various applications. Use of deep learning in music analytics along with possible advantages and word of caution is covered. Enormous future growth of music data and the related challenges with new directions are covered in the concluding session. Deep learning is a recent approach developed with neural networks for pattern recognition Deep learning terminology refers to the framework in which the entire process from the input data to the output of an application is learned from the data. This approach reduces the requirement of prior knowledge about domain and problem. It slashes overall engineering efforts and human intervention in feature generation and other tasks related to the application. Artificial neural network (NN) is a parallel computational model consists of variable interconnections of simple units to simulate human brains. The success depends on the quality of training data and the algorithms used for a specific application. The increasing acceptance of neural network models to solve PR problems have been mainly due to their capacity to learn intricate non-linear input-output relationships, little dependency on domain-specific knowledge and availability of efficient learning algorithms. Many variants of NN have been used successfully for different PR problems in music.

Deep learning attempts to model data abstraction using multiple hidden layers of the neural network. Music audio files in image format are presented as an input in the convolutional neural network. Convolutional layers attempt to abstract the high-level features from the music image. Fully connected layer attempts to map high level features abstracted to the classes and provides probabilities of different n classes shown as P1, P2, ... Pn. The class with higher probability is likely the classification of input data. Music classifications for genre, artist, instrument, emotions are different applications being attempted using a deep convolutional neural network. It replaces the need for handcrafted features using feature learning and extraction algorithms. It transforms input data through multiple layers with artificial neuron as a processing unit which are trained using algorithms such as backpropagation algorithm which is widely used.

The conventional feature engineering approach has been successfully used for various applications and it has been the favorite method by many researchers since the last few years. In the case of small data and standard computational resources available, this approach is more suitable. Deep learning requires huge data and heavy computational power for the desired results in a faster way. Training of deep learning is quite time-consuming. For some applications in which features are distinctly noticeable and captured successfully, the conventional approach is more suitable and likely to provide better results.

Automatic feature abstraction and learning using a deep learning approach is a recently developed method. Deep learning with multiple hidden layers of the artificial neural network is a promising learning tool which has shown remarkable accuracy improvements recently in the fields like speech recognition, image processing. Deep learning is more suitable for multimedia data like images, audio, video in which the features are readily not available in numbers. Music data is an excellent candidate for deep learning and researchers are exploring a new approach for music data. A hybrid approach using both handcrafted features and features from deep learning has improved the results in an application such as music recommendation. The full deployment of this approach in computational music is still awaited and we may get exciting results and applications in the near future.

The real challenge is to achieve computational intelligence in musicology is music modeling. We humans ourselves have not able to judge the entire musical aspects in the true sense to model it for further processing. Modeling human perception and fast feature learning scalable parallel algorithms will lead to further progress in the domain. An interesting end to end applications for various tasks using more advanced music recognition systems is likely to dominate in the coming years. Structural approaches along with statistical methods as a combined technique may be more useful for musical applications as music is a felt phenomenon and not only just numbers. Although quantitative evaluation dominates the present research, mixed methods with combining qualitative and quantitative analysis can be more useful for music analytics. Human perception is subjective in nature and differences in perspectives lead to limited inter-rater agreement. These levels of inter-rater agreement illustrate a natural upper bound for any sort of algorithmic approach.

Incorporation of human music perception and cognition in feature engineering will provide better results. Musical knowledge and pattern representation using advanced visualization techniques to model and simulate the human system will advance in the coming years with more insight into human music decoding. Considering millions of available music tracks and the enormous growth of music over the Internet, tasks such as music pattern analysis and retrieval for the huge growing data will be a challenge in the coming years. Numerous innovative learning algorithms can be proposed in the near future for efficient music retrieval. Modeling human perception and fast adaptive learning algorithms will be the key to designing future intelligent systems.

References

[1] Aljanaki, A., Yang, Y.H. and Soleymani, M., Developing a benchmark for emotional analysis of music. *PloS one*, 12(3), p.e0173392, 2017.
[2] Alpaydin, E. Introduction to machine learning. *MIT Press*, 2014.

[3]   Arifi, V., Clausen, M., Kurth, F. and Müller, M. Automatic synchronization of music data in score-, *MIDI-and PCM-format*, 2003.
[4]   Bhalke, D.G., Rao, C.R. and Bormane, D.S. Automatic musical instrument classification using fractional Fourier transform based-MFCC features and counter propagation neural network. *Journal of Intelligent Information Systems*, 46(3), pp.425–446, 2016.
[5]   Bisot, V., Serizel, R., Essid, S. and Richard, G. Feature learning with matrix factorization applied to acoustic scene classification. *IEEE/ACM Transactions on Audio, Speech, and Language Processing*, 25(6), pp.1216–1229, 2017.
[6]   Borgatti, S.P., Everett, M.G. and Johnson, J.C. Analyzing social networks. *Sage, 2018.*
[7]   Chandrashekar, G. and Sahin, F. A survey on feature selection methods. *Computers & Electrical Engineering*, 40(1), pp.16–28, 2014.
[8]   Datta, A.K. etl., A. Music Information Retrieval. In Signal Analysis of Hindustani Classical Music (pp. 17–33). *Springer, Singapore, 2017.*
[9]   De Prisco, R., Esposito, A., Lettieri, N., Malandrino, D., Pirozzi, D., Zaccagnino, G. and Zaccagnino, R. Music plagiarism at a glance: metrics of similarity and visualizations. In Information Visualisation (IV), *21st International Conference IEEE* (pp. 410–415), 2017.
[10]   Deshmukh, S.H., Bajaj, D.P. and Bhirud, S.G. Audio descriptive analysis of singer and musical instrument identification in north Indian classical music. *International Journal of Research in Engineering and Technology*, 4(6), pp.505–508, 2015.
[11]   Dittmar, C., Hildebrand, K.F., Gaertner, D., Winges, M., Müller, F. and Aichroth, P. August. Audio forensics meets music information retrieval—a toolbox for inspection of music plagiarism. In *Signal Processing Conference (EUSIPCO)*, Proceedings European (pp. 1249–1253). IEEE, 2012.
[12]   Grosche, P., & Müller, M. Extracting Predominant Local Pulse Information from Music Recordings. *IEEE Transactions on Audio, Speech, and Language Processing*, 19, 1688–1701, 2011.
[13]   Hu, X., Choi, K., and Downie, J.S. A framework for evaluating multimodal music mood classification. *Journal of the Association for Information Science and Technology*, 68(2), 2017.
[14]   João Lobato Oliveira; Matthew E. P. Davies; Fabien Gouyon; Luís Paulo Reis Beat Tracking for Multiple Applications A Multi-Agent System Architecture with State Recovery, *IEEE Transactions on Audio, Speech, and Language Processing*, Volume 20 Issue 10, Page 2696–2706, 2012.
[15]   Lerch, A. An introduction to audio content analysis: *Applications in signal processing and music informatics.* Wiley-IEEE Press, *2012.*
[16]   Liem, C., Müller, M., Eck, D., Tzanetakis, G. and Hanjalic, A. November. The need for music information retrieval with user-centered and multimodal strategies. In *Proceedings of the 1st international ACM workshop on Music information retrieval with user-centered and multimodal strategies* (pp. 1–6), 2011
[17]   Makarand, V. and Parag, K. Unified Algorithm for Melodic Music Similarity and Retrieval in Query by Humming. In *Intelligent Computing and Information and Communication* (pp. 373–381). Springer, Singapore, 2018.
[18]   Michalski, R.S., Carbonell, J.G., and Mitchell, T.M. eds. Machine learning: An artificial intelligence approach. *Springer Science & Business Media, 2013.*
[19]   Müllensiefen, D. and Pendzich, M. Court decisions on music plagiarism and the predictive value of similarity algorithms. *Musicae Scientiae*, 13(1_suppl), pp.257–295, 2009.
[20]   Panteli, M., Benetos, E. and Dixon, S. A review of manual and computational approaches for the study of world music corpora. *Journal of New Music Research*, 47(2), pp.176–189, 2018.
[21]   Rao, V. and Rao, P. Vocal melody extraction in the presence of pitched accompaniment in polyphonic music. *IEEE transactions on audio, speech, and language processing*, 18(8), pp.2145–2154,2010.

[22] Rosa, R.L., Rodriguez, D.Z. and Bressan, G. Music recommendation system based on the user's sentiments extracted from social networks. *IEEE Transactions on Consumer Electronics*, 61(3), pp.359–367, 2015.

[23] Salamon, J. and Gómez, E. Melody extraction from polyphonic music signals using pitch contour characteristics. *IEEE Transactions on Audio, Speech, and Language Processing*, 20(6), pp.1759–1770, 2012.

[24] Schedl, M. and Schnitzer, D. Location-aware music artist recommendation. *International Conference on Multimedia Modeling* (pp. 205–213). Springer, Cham, 2014.

[25] Serra, X. Creating research corpora for the computational study of music: the case of the Compmusic project. In Audio Engineering Society Conference:53rd *International Conference: Semantic Audio. Audio Engineering Society, 2014.*

[26] Shmulevich, I., Yli-Harja, O., Coyle, E., Povel, D.J. and Lemström, K. Perceptual issues in music pattern recognition: Complexity of rhythm and key finding. *Computers and the Humanities*, 35(1), pp.23–35, 2001.

[27] Smith, J.O. Spectral audio signal processing (Vol. 1334027739). *W3K, 2011.*

[28] Thickstun, J., Harchaoui, Z. and Kakade, S. Learning features of music from scratch. *arXiv preprint* arXiv:1611.09827, 2016.

[29] Thorat, etl. Survey on collaborative filtering, content-based filtering, and hybrid recommendation system. *International Journal of Computer Applications*, 110(4), 2015.

[30] Velankar, M., Deshpande, A. & Kulkarni, P. Melodic pattern recognition in Indian classical music for raga identification. *Springer International Journal of Information Technology, 2018.*

[31] Velankar, M.R., Sahasrabuddhe, H.V. and Kulkarni, P.A. Modeling melody similarity using music synthesis and perception. *Procedia Computer Science*, 45, pp.728–735, 2015.

[32] Wang, X. and Wang, Y. Improving content-based and hybrid music recommendation using deep learning. In Proceedings of *ACM international conference on Multimedia* (pp. 627–636), 2014.

[33] www.ismir.net accessed 4 Dec 2018

[34] www.smcnetwork.org/software.html accessed 4 Dec 2018

Mamatarani Das, Mrutyunjaya Panda, and Shreela Dash

4 A Comparative Analysis of Machine Learning Techniques for Odia Character Recognition

Abstract: Character recognition is a challenging area in Machine Learning, Pattern Recognition or Image Processing. The accuracy to recognize handwritten character by human is far better compared to machine recognition. To develop an interface which can differentiate characters written by human yet requires intensive research. Though number of researches have presented in this area, still research is going on to achieve human like accuracy. Both handwritten and printed character recognition are categorized into two types, online and offline. A good number of researches have done work in the area of optical character recognition in different languages but for the Odia language, development is negligible. Odia (formerly it was Oriya), one of the 22 scheduled language recognized by the constitution of India and it is the official language of the state of Odisha (Orissa), more than 40 million people speak Odia. Due to the roundish shape of Odia character, large number of modified and compound characters and similarity between different characters makes this language very hard to create a satisfactory classifier. In the present survey undertaken we have discussed what are challenges be for Odia language and the machine learning techniques used in the recognition of Odia character recognition. This chapter describes complete process of character recognition i.e. pre-processing, extraction and selection of feature set and character recognition elaborately with comparison analysis and the metrics used to evaluate machine learning algorithms.

Keywords: Odia language, Optical Character Recognition (OCR), Machine Learning, Handwritten Character Recognition, Printed Character Recognition, Evaluation Metrics.

4.1 Introduction

Language helps us to express our thoughts. India is a multi-lingual, multi-script country. Ten official Indian scripts are Devnagari, Gurmukhi, Bangla, Tamil, Telgu, Urdu, Odia, Gujarati, Kannada and Malayalam. Most of them are originated from Brahmi. In recent times, there have been efforts made in automation and character recognition for all Indian languages, which will not only enhance technical use of the knowledge, but also preserve the language for next generation and showcase the development to its researchers. Optical Character Recognition, i.e., considering OCR, electronically converts images from printed, handwritten or typewritten texts to digital version or the images of characters into a standard encoding scheme like ASCII or

https://doi.org/10.1515/9783110610987-006

Unicode to represent them. Presently, OCR technology is being used in large number of the industries for better record keeping of documents.

In this chapter we discuss OCR for Odia language, which is not only different from European language, but also its script is very different from its sister languages like Bengali and Assammese. The major difficulties in distinguishing between numerals or characters are by their font sizes and variation in writing styles also. This character recognition with due level of difficulty is divided into online and offline group.

If the characters are identified from already written character patterns that are scanned and stored as digital images then it is known as off-line character recognition whereas in the online one the document or character get processed as soon as it is generated. On-line character recognition identifies the characters while they are written and it deals with sequences of strokes of pen movement in time domain like pen up and pen down and pads that are highly receptive to pressure that record the pressure and velocity of pen's movement. The offline process oblivious to external pressure, even pace of writing, but online is definitely gets affected. Thus offline is more applicable to handwritten as well as printed documents and the online more suited to optical characters. In online the raw data is acquired in samples/second whereas in off-line documents, the texts or characters are digitized to dots/inch. Real time contextual information is associated with the processing of online data and the recognition and accuracy rate of online character recognition is higher compared to offline recognition. However both types of character recognition are a stepwise process starting from generation, digitization, storing in memory, feature selection and finally recognizing the character. If we compare the flow graph with English or any other language's character recognition, the same process is followed for Odia character recognition.

Off-line character recognition, especially handwritten character recognition is more complex compared to online recognition process and needs more investigation in view of online character recognition. Due to paucity of research materials for Odia languages, we try to find out present trends in Odia character recognition, prevailing methods, limitations are analyzed with scope for research is being suggested.

The rest of the chapter is presented as on Section 4.2 describes the overview of character recognition and their types, Odia character set and their characteristics are discussed in Section 4.3, standard databases for Odia character recognition is discussed in Section 4.4, in Section 4.5, recognition of offline Odia script as well as the comparison of different machine learning techniques used so far for classification and recognition of printed as well as handwritten alphabets and numerals are discussed, Section 4.6 describes different types of evaluation metrics used for machine learning algorithms, Section 4.7 describes the challenges of Odia text recognition and Section 4.8 is concluding section.

4.2 Overview of Character Recognition

Optical character recognition (Character recognition) is a field of computer science which has started its development as early as 1950. The first step of any character recognition is to know the basics of different characters for that language. Every researcher has to go back to the roots of the language and knowing thoroughly about all the characters which are required for making up the language.

Broad applications of character recognition are,

1. To take classroom note or to enter text into computer automatically
2. Word processing
3. Producing annotated hand sketch
4. Helping blind people to read at a great extent
5. To fill a form by digital pen
6. Preserving the historical document for next generation
7. Bank cheque processing
8. Applied to cell phone which is related to handwriting recognition
9. Automatic processing and sorting of postal letters by PIN code
10. Read and manipulate the text in any printed document and many more.

Broadly, the character recognition is classified into two categories, online and offline.

4.2.1 Online Character Recognition

Due to tremendous growth of technology in our day to day life, the involvement of portable computing devices like personal digital assistants or any kind of handheld computers, where non-keyboard methods are providing a better way for man machine interface. Most of the today's data is entered through keyboard, many tasks still exist where people prefer the handwritten input over keyboard entry or voice based input. The act of writing through a special pen on an electronic surface and this online handwriting is recognized by keeping the record of hand movements is the challenge to online character recognition. So online character recognition provides a much better data entry method where keyboard entry is a very difficult task.

The process of recognition of online handwritten characters includes following phases: data acquisition, pre-processing, line, word and character segmentation, feature extraction, classification and recognition and post-processing. A brief outline of these phases for online character recognition is described below.

Data Acquisition: Online handwritten recognition requires a device that captures writing as it is written. Electronic tablet or any kind of digitizer is used mostly to capture and a digital pen is used for this purpose. Most common parameters used for data acquisition are pen tip position, acceleration, velocity and sometimes pressure on the writing surface.

Pre-processing: In this phase of handwriting recognition, noise or distortions that are present in input text due to hardware and software limitations are removed. A number of operations are carried out in pre-processing to improve the quality of characters and make them suitable for the segmentation of lines, words or characters. Typical steps followed by pre-processing are noise removal, binarization, skew angle correction, resizing the image, normalization, smoothing, resampling, thinning etc. to enhance the quality of scanned image.

Segmentation: The pre-processed image is taken for line, word and character segmentation. Row histogram is used for line segmentation. All words of a row are extracted using column histogram. Then the characters are extracted from words. However in the case of online, as the written characters are associated with temporal information, so any overlapping characters will not pose any problem, as they are still different by the time difference of the characters or text written. Accuracy of character recognition highly depends on accuracy of segmentation.

Feature Extraction: The process of collecting distinguishable information for a character or group of characters and on the basis of this information we can classify to a class is called feature extraction. The choice of feature set and extraction of features play a very important role in the design of successful OCR system. The relevant information required for further processing is extracted from the input. Typically shape, directional, length, angle features are extracted. The challenge in this phase is to find a minimal set of features with maximum data recognition.

Classification and Recognition: The reduced feature set obtained from feature extraction phase is input to different classifiers. Methods like SVM (Support Vector Machine), HMM (Hidden Markov Model), NN (Neural Networks) are frequently used for online handwriting recognition. The aim is to find the optimal letter for a given online handwritten input.

Post processing: To improve the performance of the system post processing is performed. It includes the procedure of correcting misclassified results.

If we look into the history of online Odia character recognition, to the best of our knowledge, there is no published work on online character recognition for Odia script. So we cannot compare results for this online recognition of characters. However number of works had done for other Indian language in this branch of research. In this paper we have analyzed and compared different techniques used for offline i.e. both printed and handwritten recognition used for Odia character.

4.2.2 Offline Character Recognition

Characters are recognized offline by two different ways, printed document recognition and handwritten character recognition. Handwritten Character Recognition (HCR) is a very complex task due to different writing styles and variations of handwriting for different individuals that can produce extreme differences in characters shapes. Whereas the printed document recognition is comparatively easier than

HCR as the system has to work on same font type. If we see back the history of OCR for Odia language, [1][2] had done the foundation step to develop a system for Odia character recognition for printed characters. Like online character recognition, most current approaches to offline character recognition consists of main stages namely data acquisition, pre-processing, feature extraction, recognition by classification techniques and post processing. A sample of OCR for offline character recognition is given in figure 4.1.

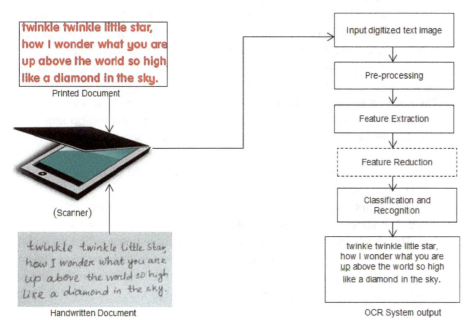

Fig.4.1: Sample of printed and handwritten text recognition

4.3 Odia Character Set and Their Characteristics

Odia is a most popular Indo-Aryan language and is spoken by people of Odisha as well as the neighbors like Chhatishgarh, Andhrapradesh, West Bengal and Jharkhand. Odia script is derived from Brahmi script of ancient India. Basically Indian language alphabets are systematically arranged from the phonetic point of view. Vowels come first in which the long ones immediate it follow their short counterparts and then the consonants are grouped according to their places and the manners of articulation. Modern Odia language has total 47 letters comprising 11 vowels and 36 consonants if we omit anuswara, bisarga and chandrabindu.

The characteristics of alphabets and numerals are:
1. The writing style of Odia character is unique as compared to any other languages of India. Most of its characters are roundish in nature because in ancient times

people used palm leaves with pointed tip pen for writing and due to the circular shape, which has a tendency to tear for horizontal or vertical strokes.

2. There is no line at the top of a character such as the script of Devnagari.
3. Apart from these 47 letters and 10 numerals, also by merging a vowel with consonant or combining two consonants composite characters are formed.
4. Along with consonants and vowels there are some special symbols namely anuswara, bisarga, chandrabindu. These symbols are also combined to form another composite character.
5. Vowels are shown as an independent vowel or can be found in any part of a word or can be attached to upper, lower, left or right-hand side of consonants to form them as modifiers. A consonant is combined with another consonant to form compound characters.
6. Odia language is not case sensitive.
7. The end of a sentence is marked by a vertical line ('|'), not by a dot ('.'). But, a few characters like 'ଆ' or any consonant + ଆ (i. e. �) have a vertical line at the end. Due to the presence of this vertical line, character segmentation is very difficult.
8. Some characters have similar type of structure like digit '0' (Zero) and the alphabet 'O' ('tha') are similar in shape.
9. The Odia script text is divided in the upper, middle and lower areas known as zones. The upper zone is made up of the portion between the mean and upper line. The area below the mean line and above the base line is called the middle zone and the part between the base and the lower line is the lower zone. The line between the middle and lower zones is known as the base line and the line between the upper and middle zone is called the mean line.

4.4 Database for Odia Language

Any fruitful work in the area of text recognition needs a benchmark database. Large databases are required for robust training of the recognizer or classifier. Due to cursive scripts and different styles of handwriting, the accuracy of recognition fully dependent on the type of feature extractor used and the amount of training samples from the database.

In 2005, an isolated handwritten Odia numeral database was prepared by [3], at ISI Kolkata, India. Exactly 356 people were involved in the data collection process. It contains 5,970 samples collected from 166 application forms, 105 pieces of mail and rests collected personally. Finally, the data set is divided into a training set and a test set, which includes 4,970 samples and 1,000 samples. Compared with the Bangla digits data set size, the size of the Odia digit database is too small.

There are few other data sets, in which researchers have evaluated their performance of their classification algorithms and reported in the literature, but they are not available to the public and there is also no clear mention of the sample size or how the database was prepared.

The authors of [4] discussed a new database for Odia numerals and characters at IIT Bhubaneswar. The material was scanned at 300 and 600 dpi. There are currently 5,000 handwritten samples of Odia numerals and 35,000 handwritten character samples are in the IITBBS database. The number of classes in the handwritten number database of IITBBS is 10, while the number of classes in the handwritten Odia database of IITBBS is 70. On request, these databases are freely available to the users.

At NIT Rourkela, also databases are prepared and described by [5], which includes both Odia alphabet and digit with a total of 18,240 samples. The database segregated into two sets namely Odia Digit Database (ODDB) and Odia Handwritten Character Set (OHCS). The ODDB database initially contains 3,200 digit image samples, which rose to 9,000 after collecting 580 more samples per digit. The OHCS database contains 15,040 samples of atomic characters from Odia language. The steps adopted in the process of designing the database on Odia character set are data acquisition, image enhancement, and size normalization and store it in database.

4.5 Recognition of Offline Odia Script

Figure 4.2 represents the standard workflow that is maintained for Odia character recognition. In the first phase of recognition process, different image processing steps that are involved in pre-processing of character images are,

$$\text{RGB image} \rightarrow \text{Gray Scale Image} \rightarrow \text{Binarization} \rightarrow \text{Inversion}$$

The text document is scanned by a scanner and stored as gray scale image in 200/300 dpi, where the intensity values of image lies between $0-255$. Then the binarization of image converts gray scale image which have only pure white and black pixels. Then it is subjected to inversion process where each pixel gets an inverted color. Then it is subjected to pre-processing step, as in the process of text digitization, some unavoidable situations like the document may contain noise while scanning or even it may be skewed.

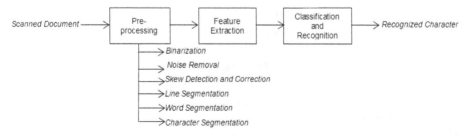

Fig. 4.2: Standard workflow of character recognition system

The number and types of pre-processing algorithm that is applied to a scanned image depends on several factors like, age of the document, scanned image resolution, quality of paper / material, amount of skew in the image, image and text format, type of the script used like handwritten or printed etc..

The better the pre-processing step, the better the feature extraction process, which indirectly dependent on the quality of classification process. Character features represent both the morphological and the spatial properties. Extraction of features is a process of extracting dominant features of a character that are extremely used for classification. Statistical feature extraction, structural feature extraction and image transformations are basically three types of feature extraction techniques.

The extracted function vector set is the combination of all the functions extracted from each character in statistical extraction [6]. The associated feature in the feature vector is due to the relative position of the features in the character image matrix in this type of extraction. Morphological features of an image matrix character such as edges, curvature, regions etc. by structural feature extraction method extracted. This method extracts the character features according to how characters are written.

Popular image transformations used to represent the feature vector are Fourier transform, Stockwell transform [7], Wavelet transform [8], Slantlet transform [7]. Features of an Odia character are collected after character segmentation process of pre-processing step, it is necessary to find best feature subset that are required for classification. Stroke and run-number based features and the features obtained from the concept of water reservoir [1], binary external symmetry axis constellation [9], dimensional features like height and width, centroid, histogram pixel count in row, column and total pixel [10], zone pixel [11], features from chain coding method [12], binary and structural features [13], curvature feature i.e. features of concave, linear and convex region and gradient feature [14], Stockwell transform [15] etc. are distinct features that are taken by different authors. Feature selection is the process to find the best subset of features from the input set. Offline character recognition is more challenging than online character recognition and in offline case the handwritten recognition process is complex than printed document recognition, because there are no generic features that are suitable for all types of classification algorithms.

The classification and character recognition system are the decision making part, where the overall performance of the system depends greatly on the type of classifier used. The extracted features after feature extraction stage, are input to a classifier and for a given input (test samples), the classifier compares it with the stored features (training samples) to assign a class for the input.

Traditionally, template based approach and feature based approach are used as the process of character classification and recognition. In a template-based approach, the matching character is matched to the ideal template pattern stored and the similarity between the two is used for the classification decision. The authors of [16] used the structural features of characters and employs template matching technique for recognition. The character recognition system had employed earlier this template matching approach, but is not so effective in the presence of noise or different styles of handwriting. Modern systems combine it with feature based approach to get better result.

In feature based approach, from a character the important properties or features are obtained from test patterns and apply them to a classification model. Different recognition classifiers that used by the authors of Odia character recognition are decision tree [1], feed forward neural network [10], finite automata [12], Tesseract OCR [28], back propagation neural network [13], genetic algorithm optimization neural network(GAONN) [17], multiclass SVM [18], random forest [9], k-Nearest neighbor [9], quadratic classifier [14], HMM [19], DLQDF [20], Hopfield NN [21], ANN [22], Naïve Bayes classifier [23], MQC [24], recurrent neural network(Elman Network) [25].

4.5.1 Different Machine Learning Classification Algorithms for Character Recognition

In the field of computer science, machine learning can be defined as an area which defines the computers as the ability to learn without being explicitly programmed. Supervised learning and unsupervised learning are the types of learning methods used in neural networks. To train the network, the target output pattern is associated with every input pattern. A comparison is made between the system's computed output and the expected output. Error is computed and is used to adjust the weights of the network. Regression and classification are two types of supervised learning. In unsupervised learning, the target output and a mentor is not present in the system. The system learns by its own to adapt the input patterns. Clustering technique is basically used for unsupervised learning.

Classification is the task of allocating an unknown object to one of the predefined categories is shown in figure 4.3.

Fig. 4.3: Task of a classifier

Different machine learning classification techniques used are:
- Decision based learning
  - Decision tree
- Artificial neural network
  - Multilayer network and Back propagation algorithm
- Bayesian Learning
  - Naïve- Bayes classifier
  - Bayesian belief network

- Instance Based Learning
  - k-nearest neighbor learning
- Genetic algorithm

4.5.1.1 Decision Tree

Decision tree learning is one of the widely used methods for any type of character recognition. Decision tree is used to represent learned function. The authors of [1] presented the optical character recognition on printed Odia script, where the digitized image is pre-processed by several modules like skew correction, zone detection, line, word and character segmentation etc. The characters are recognized by a decision tree based classifier and it uses the features obtained from the concept of water flow from a reservoir along with the stroke and run number based features, provides an accuracy of 96.3%. Some of the structural features used like circular upper part of character , vertical line on right most part, horizontal or vertical line code, number and position of holes etc. In this approach, the similar structure characters are grouped together and then classified based on the feature vectors. This method is not so strong to capture the characters of similar structures.

4.5.1.2 Artificial Neural Network (ANN)

The Artificial Neural Network is a collection of many interconnected cells. Cells are arranged so that each cell receives an input and produces an output for subsequent cells. Figure 4.4 shows the structural block diagram and work flow of the artificial neural network and figure 4.5 shows the multi-layer feed forward neural network. The neurons in the neural network are serially interconnected. In the ANN system there are one input layer, more than one output layer and some intermediate layers. There are a number of layers in the network, while all other intermediate layers are hidden except the last output layer. The input layer output is supplied to the hidden layers and the output from the hidden layers is then supplied to the output layer.

To train the system efficiently, back propagation learning method can be applied to the multilayer feed forward network. It is one of the supervised learning algorithms and is a two-step process.

Propagation: Forward propagation of training pattern's input is passed through the neural network to generate the propagations output activities. Error is calculated between the network output and target output which is the difference between the actual output of each neuron and its desired value as given for the input pattern taken from the training set. Mathematically the error (E) is presented in equation (4.1).

$$E = 12 \sum_{j} (d_j - O_j)^2 \qquad (4.1)$$

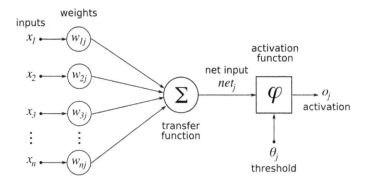

Fig. 4.4: Basic structure and workflow of an ANN

Where, for some input vector d_j represents the desired output and O_j represents the actual output of the jth output neurons.

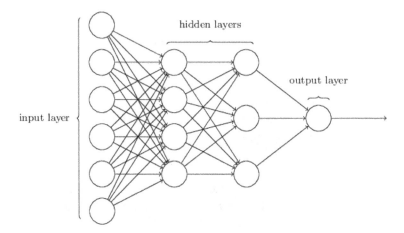

Fig. 4.5: Structure and workflow of a multilayer feed forward ANN

Weight Update: Error that is generated after computation is propagated backwards through the network to adjust the weight and is shown in in figure 4.6.

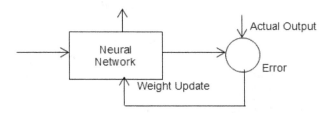

Fig. 4.6: Block diagram of Back propagation neural network

In [22], the authors used feed forward, multi-layered neural network, that is used for the recognition of Odia numerals. To train the neural network error Back-propagation (EBP) algorithm is used. The features that is the input for the classifier is the mean of the object pixels for different quadrants, to identify the required numerals. The overall handwritten numeral recognition accuracy of their scheme is about 93.20 % with a rejection rate of 6.8 %.

4.5.1.3 Naïve-Bayes Classifier

Naïve Bayes classifiers are very well known as probabilistic classifier, based on applying Bayes' theorem. It assumes a strong independence among the features. The advantage of this type of classifier is the accuracy of classification result with a minimum training time as compared to other unsupervised or supervised learning algorithms. For text or pattern recognition, mathematical expression for Bayes' theorem is given in equation (4.2).

$$P(w_j|x) = P(x|w_j).P(w_j)\sum_{k=1}^{N} P(x|w_k).P(w_k) = \left[P(x|w_j).P(w_j)P(x)\right] \tag{4.2}$$

Where

$P (w_j)$ = Prior probability of Class w_j
$P(w_j|x)$ = Posterior probability of class w_j, given the observation x
$P(x|w_j)$ = Likelihood (conditional probability of x given class w_j)
$P(x)$ = A normalization constant

The authors of [23], described a Naïve-Bayes algorithm for recognition of handwritten Odia numerals of ISI Kolkata numeral dataset by taking LU factorization technique for extraction of features having average accuracy of 92.75 %.

4.5.1.4 k-Nearest Neighbor Learning

The k-nearest neighbor algorithm is one of the basic classification algorithms and is popular due to its simplicity and good performance. Figure 4.7 represents the example of k-nearest learning method.

In k-NN algorithm, the nearest training patterns for each test pattern are searched and the most common classes are taken and assigned to the test pattern. 'k' therefore represents the number of training data points closer to the test data point for which we going to find the class.

Limitations of this method are: (i) It stores the entire training set; (ii) Search the entire training set in order to classify a given pattern; (iii) The performance of classification is degraded on the presence of noisy data.

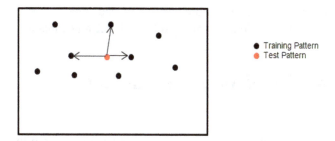

Fig. 4.7: Structure of k – nearest learning

The authors of [9][5], have used k-nearest classification by using binary external symmetry axis constellation and Stockwell transform features for classification of Odia characters and got an accuracy of more than 98 % over IITBBS and ISI Kolkata numeral database.

4.5.1.5 Genetic Algorithm

Genetic algorithm uses chromosome and collection of chromosomes are referred to as population. A chromosome is comprised of smallest unit called genes and the value of a gene can be numerical, binary, symbols or characters. The fitness functions are applied over these chromosomes to find the solution generated by GA with the given problem. Some chromosomes in population will passed through process of crossover and producing new chromosomes. A few chromosomes are applied for the mutation process in their gene. Finally the value of chromosome will remain to a certain value and this value is considered as the best solution for the problem.

Basic steps of genetic algorithm are to initialize the population, selection of chromosomes according to fitness function, crossover between selected chromosomes, perform mutation and repeat the cycle to find a best solution.

4.5.2 Comparison of different techniques used for printed as well as handwritten character recognition

To the best of our knowledge, a few numbers of publications are published in printed Odia script. In the literature review it shows that mostly neural network has been chosen as classifiers. Combinations of artificial neural networks and genetic algorithms have been reported in [26], provides also satisfactory results. Nearest neighbor classifiers are better compared to SVM as they occupy minimum storage space and take less computation time. Most of the authors have taken IIT Kolkata numeral database, IITBBS numeral and character database or NIL Rourkela database or their own created database that is not mentioned properly. Table 4.1 shows different types of techniques of used for printed Odia character recognition. A number of works had done in case of handwritten character and digit recognition, summarized in Table 4.2 and Table 4.3.

4.6 Evaluation Metrics of Machine Learning Algorithms

From Table 4.1, 4.2 and 4.3, we have seen different classification algorithms used to classify printed as well as handwritten Odia characters and their recognition accuracy. Accuracy is a measure to determine how close we are coming to the target result. Accuracy is the most commonly used evaluation metric to assess how good a machine learning algorithm and its model is to take a test. We usually mean the classification accuracy as the ratio of number of correctly classified samples or instances or points to the number of total samples given to the classifier and is given in equation (4.3).

$$Classification\ Accuracy = \frac{Total\ number\ of\ correct\ predictions}{Total\ number\ of\ samples\ given} \qquad (4.3)$$

This accuracy method works well if the number of samples of each class is same. But a high accuracy model need not be a good model. Accuracy is not sufficient to describe how good a model is. In machine learning we deal with two types of problem i.e. classification and regression problem and based on those problems, different evaluation metrics used are classification metrics and regression metrics. Confusion matrix, accuracy, precision, recall, F – measure, Receiver Operating Characteristics Curve (ROC Curve) are used under classification metrics and whereas Mean Absolute Error (MAE), Mean Square Error (MSE), R2 score are used under regression metrics.

Confusion Matrix: It is a table that describes the complete performance of the model. This measure is defined by *true positive (TP), false positive (FP), true negative (TN), false negative (FN)*. Confusion matrix is shown in Table 4.4.

Tab. 4.4: Confusion Matrix

Actual / Predicted	Positive	Negative
Positive	True Positive(TP)	False Negative(FN)
Negative	False Positive(FP)	True Negative(TN)

True Positive (TP): Positive samples that are classified as positive by the model
- True Negative (TN): Negative samples that are classified as negative by the model
- False Negative (FN): Positive samples, but model incorrectly classified it as negative
- False Positive (FP): Negative samples but model incorrectly classified it as positive

Tab. 4.1: Recognition of printed Odia characters and their recognition accuracies

Author	Input	Pre-processing techniques	Features	Recognition Classifier	Data Base	Data samples	Recognition accuracy
B.B.Chaudhury, U.Pal, M.Mitra [1]	Printed Odia character	Skew detection and correction, line, word and character segmentation	Stroke and run-number based features along with the features obtained from the concept of water reservoir.	Decision tree	ISI Kolkata	-	96.3
S.Mohanty, H.N.D.Bebarta [27]	Printed Odia document	Binarization, Skew correction by Baird algorithm	NA	Decision tree based SVM	NA	10000	Depends on font type and size, Average is 96%
S. Mishra, D. Nanda, S. Mohanty [10]	Printed Odia Character	Binarization, Thinning and edge detection	Dimensional features like height and width, centroid, histogram pixel count in row, column and total pixel	Feed forward neural network			
S.Mohanty, H.N.D.Bebarta, T.K.Behera [11]	Printed Odia Document	Noise removal, Skew detection and correction, line, word, and character segmentation by horizontal and vertical projection profile	Zone pixel	Structural Analysis			
R.K.Mohaparta, B.Majhi, S.K.Jena [12]	Printed Odia digit	Binarization, Skeletonization by chain coding, noise removal, segmentation	Features from chain coding method	Finite Automata	Odia Digit Database, NIT Rkl		96.08%
M.Nayak, A.K.Nayak [28]	Printed Odia text document			Tesseract OCR			Depends on font size-100%
M.Nayak, A.K.Nayak [17]	Printed Odia con-	Binarization	Structural features	Back propagation neural network, Genetic al-		30samples for each	93.95 95.9

Tab. 4.1: Recognition of printed Odia characters and their recognition accuracies *(Continued)*

Author	Input	Pre-processing techniques	Features	Recognition Classifier	Data Base	Data samples	Recognition accuracy
	junct character			gorithm optimization neural network(GAONN)		conjunct character	
M.Nayak, A.K.Nayak [13]	Printed Odia character	Noise removal, Segmentation to extract character	Binary and structural features	Back propagation neural network			
C.Mitra, A.K.Pujari [18]	Printed Odia character	Directional decomposition of image by zoning	Directional, 20 pixels on 4 directions by 5 fixed zones	Multiclass SVM			95%

Tab. 4.2: Summary of handwritten Odia alphabet recognition

Author	Pre-processing techniques	Features	Data Base	Data samples	Recognition Classifier	Recognition accuracy
U.Pal, T.Wakabayashi,F.-Kimura [14]	Normalization, Segmentation	Curvature feature i. e. features of concave, linear and convex region and gradient feature. PCA is used for feature reduction		18190	Quadratic classifier	94.6
D.Basa , S.Meher [16]	Thresholding, noise reduction, segmentation	Structural features			Template matching	
K.S.Dash, N.Puhan, G.Panda [9]		Binary External Symmetry Axis Constellation (BESAC)	IITBBS Odia character database	7800	Random forest SVM K-Nearest Neighbor	89.92 93.77 95.01
I.Rushiraj, S.Kundu, B.Ray [29]	Thresholding, noise reduction, segmentation	Shadow, centroid and distance features		720	Weighted Euclidean Distance	87.6
Debananda Padhi [26]	Binarization, Noise removal, Slant/Skew correction, normalization, thinning	Zone centroid distance and standard deviation		20	Back Propagation NN with Genetic Algorithm	94

Tab. 4.3: Summary of handwritten Odia digit recognition

Author	Pre-processing techniques	Features	Data Base	Data samples	Recognition Classifier	Recognition accuracy
K.S.Dash, N.Puhan, G.Panda [9]		Binary External Symmetry Axis Constellation (BESAC)	IITBBS numeral database	5000	Random forest SVM k-Nearest neighbor	97.30 98.56 98.90
K.S.Dash, N.Puhan, G.Panda [9]		Binary External Symmetry Axis Constellation (BESAC)	ISI Kolkata numeral database	20000	Random forest SVM k-Nearest neighbor	98.44 99.02 99.35
C.Mitra, A.K.Pujari [18]	Segmentation by connected component method, normalization	Directional features by zoning method			Multiclass SVM	95.00
T.K.Bhowmik, S.K.Pujari, U.Bhattacharya, B.Shaw [30]	Binarization, strokes are extracted from binarized image	Scalar and stroke information	ISI Kolkata numeral database	5970	HMM	90.50
P.Sarangi, P.Ahmed [31]	Image cropping and resizing, binarization, Thinning	LU decomposition of matrix factors	ISI Kolkata numeral database	3300	Back propagation ANN model	85.30
B.Majhi, J.Satapathy, M.Rout [32]	Normalization, Mean filtering to obtain gray scale image	Gradient , curvature feature, Feature reduction using PCA	ISI Kolkata numeral database	500	Low complexity neural classifier	98(Gradient feature) 94(curvature)
N.Tripathy, M.Panda and U. Pal [33]	Digitization, Noise removal	Features from water reservoir concept and topological and structural features		3550		97.74
K.Roy, T.Pal, U.Pal, F.Kimura [34]	Global binarization, Noise removal	Histogram of direction chain code from contour of numerals		3850	Neural Network Quadratic classifier	90.30 94.81
K.Dash, N.B.Puhan, G.Panda [20]	Ostu's global binarization	Hybrid gradient + curvature), Feature reduction by PCA	ISI Kolkata numeral database		DLQDF	98.50
K.Dash, N.B.Puhan, G.Panda [15]	binarization, size normalisation	Stockwell transform	ISI Kolkata numeral database	500	k-Nearest neighbor	98.40

Tab. 4.3: Summary of handwritten Odia digit recognition *(Continued)*

Author	Pre-processing techniques	Features	Data Base	Data samples	Recognition Classifier	Recognition accuracy
R.K.Mohapatra, B.Majhi, S.K.Jena [12]	Binarization, noise removal, segmentation, skeletonization	Number of end point, T-joint, X-joint, loop by chain coding method	NIT Rourkela ODDB		Finite Automata	96.08
T.K.Mishra, B.Majhi, P.Sa, S.Panda [19]	Binarization, noise removal, slant correction	Shape contour of numerals		2500	HMM	96.30
T.K.Mishra, B.Majhi, S.Panda [8]	Binarization, morphological operation, Thinning, Dialation,	DCT and DWT coefficients	NIT Rourkela ODDB	500	Back propagation neural network	DCT – 92.00 DWT – 87.50
K.Dash, N.B.Puhan, G.Panda [6]	Noise removal and connected component analysis	Slantlet and stockwell	ISI Kolkata numeral database	5638	k-NN classifier	Slantlet-95.04 Stockwell-98.80
T.K.Mishra, B.Majhi, R.Dash [35]	Binarization, noise removal, standardization, normalization	Contour descriptor			Back propagation neural network	96.25
M.K.Mahato, A.Kumari, S.C.Panigrahi [22]	Digitization, normalization, Segmentation	Quadrant-Mean		16000	ANN	93.20
P.K.Sarangi, A.K.Sahoo, P.Ahemad [21]	Binarization, Noise removal	Binary Image		1790	Hopfield NN	98.40
Pradeepta K. Sarangi, P. Ahmed, Kiran K. Ravulakollu [23]		LU factorization	ISI Kolkata numeral database Created a new database for numerals	5970 7270	Naïve Bayes classifier	92.75
U.Pal, T.Wakabayashi, N.Sharma, F.Kimura [24]	Data acquisition, Binarization, Normalization	Directional information	ISI Kolkata numeral database	5638	MQC	98.40

Tab. 4.3: Summary of handwritten Odia digit recognition *(Continued)*

Author	Pre-processing techniques	Features	Data Base	Data samples	Recognition Classifier	Recognition accuracy
Pradeepta K. Sarangi, Kiran K. Ravulakollu [25]	Binarization	Calculation of row wise decimal value and scale down the decimal values		1500	Recurrent Neural Network(Elman Network)	92.41

Classification Accuracy: It is the ratio of correctly classified samples to the total number samples and is shown in equation (4.4).

$$Classification\ Accuracy = \frac{(TP + TN)}{(TP + TN + FP + FN)} \tag{4.4}$$

Precision: It is the ratio of correctly classified positive samples (TP) to the total number of positive predicted samples (TP + FP) and is shown in equation (4.5).

$$Precision = \frac{TP}{(TP + FP)} \tag{4.5}$$

Recall: It is the ratio of correctly classified positive samples to the total of true positive and false negative samples and is shown in equation (4.6).

$$Recall = \frac{TP}{(TP + FN)} \tag{4.6}$$

F-measure: It represents the harmonic mean of recall and precision obtained from the confusion matrix. Its value lies in the range of [0, 1]. It signifies the robustness and correctness of a classifier. The greater the F-measure score better is the classifier. The formula to calculate F-measure is shown in equation (4.7).

$$F - measure = 2 * \frac{(Precision * Recall)}{(Precision + Recall)} \tag{4.7}$$

4.6.1 Analysis of Classifier Evaluation

In [21], the authors have made a model using Hopfield neural network to classify handwritten Odia numerals. The total number of samples taken for numeral database is 1480 and for each class it is 148. In this implementation they got 95.4% as recognition accuracy, where the confusion matrix is given in Table 4.5.

Tab. 4.5: Sample Confusion matrix of Hopfield NN classifier [21]

Input Character	0	1	2	3	4	5	6	7	8	9	Erroneous Output
0	142	2	0	0	0	0	0	0	0	0	4
1	2	142	0	0	0	0	0	0	0	0	4
2	0	0	141	0	0	0	4	0	0	0	3
3	0	0	0	143	0	0	0	0	0	0	5
4	3	0	0	0	143	0	0	0	0	0	2
5	0	0	0	0	4	139	0	0	0	0	5
6	0	0	2	0	0	0	139	0	0	0	7
7	0	0	4	0	0	0	1	138	0	0	5
8	0	0	0	0	0	0	0	0	141	2	5
9	0	0	0	0	0	0	0	0	2	141	5

It was found that there were some cases where the proper class is not classified by the network for the given input character. The network mis-classified these characters for these cases. Some examples of confusing numerals like Odia '2' was recognized as Odia '6', Odia '5' was recognized as Odia '4' and Odia '7' was recognized as Odia '2'.

The authors of [6], compare k-NN classifier by taking the Stockwell and Slantlet transform features and tested it to the ISI Kolkata numeral data set. Each class of 500 images, randomly 400 images are taken for training purpose and 100 images are taken for test purpose and the confusion matrix is shown in Table 4.6.

Tab. 4.6: Confusion matrix of k-NN classifier [6]

Stockwell Transform Feature + k-NN Classifier											Slantlet Transform Feature + k-NN Classifier									
0	1	2	3	4	5	6	7	8	9		0	1	2	3	4	5	6	7	8	9
100	0	0	0	0	0	0	0	0	0		100	0	0	0	1	0	0	0	0	0
0	100	0	0	0	0	0	0	0	0		0	99	0	0	0	1	0	0	0	1
0	0	99	0	0	0	0	1	0	0		0	0	97	0	0	0	1	3	0	0
0	0	0	100	0	0	0	0	0	0		0	0	0	100	0	0	0	0	0	0
0	0	0	0	100	0	0	0	0	0		0	0	0	0	99	0	0	0	0	0
0	0	0	0	0	100	0	0	0	0		0	0	0	0	0	99	0	0	0	0
0	0	1	0	0	0	100	0	0	0		0	0	1	0	0	0	98	1	0	0
0	0	0	0	0	0	0	99	0	0		0	0	2	0	0	0	1	96	0	0
0	0	0	0	0	0	0	0	100	0		0	0	0	0	0	0	0	0	100	1
0	0	0	0	0	0	0	0	0	100		0	1	0	0	0	0	0	0	0	98

The authors also compare their methods with [30], [34], [24], [20], with different classifiers like HMM, Quadratic, MQDF, DLQDF and k-NN and found the accuracy rate of k-NN classifier is 98.8% and error rate is 1.2% (Table 4.7).

Tab. 4.7: Classifier accuracy

Author	Feature	Classifier	Accuracy (%)	Error (%)
T.K.Bhowmik, S.K.Pujari, U.Bhattacharya, B.Shaw [30]	Scalar	HMM	90.50	9.50
K.Roy, T.Pal, U.Pal, F.Kimura [34]	Directional	Quadratic	94.81	5.19
K.Dash, N.B.Puhan, G.Panda [6]	Slantlet	k-NN	95.04	4.96
U.Pal, T.Wakabayashi, N.Sharma, F.Kimura [24]	Directional	MQDF	98.40	1.60
K.Dash, N.B.Puhan, G.Panda [20]	Hybrid	DLQDF	98.50	1.50
K.Dash, N.B.Puhan, G.Panda [6]	Stockwell	k-NN	98.80	1.20

In [34], the authors applied the Neural Network and the Quadratic Classifier to 3,850 numerals obtained from 385 different individuals with 64 and 400 dimensional characteristics and noted the overall accuracy of recognition as 90.38% with a refusal rate of 1.84%. The confusion pair of numerals and the error vs. rejection rate obtained from the neural network and quadratic classifier is shown in Table 4.8.

Tab. 4.8a: Confusing Odia numeral [34]

Tab. 4.8b: Error rate vs. rejection rate of classifier [34]

Sl. No.	Numeral Class	Classified as Numeral Class	NN 64 feature	Quadratic classifier	
				64 feature	400 feature
1.	6	7	1.35	2.38	1.24
2.	7	2	0.92	1.00	0.67
3.	6	2	0.27	1.24	0.97
4.	2	7	0.54	0.70	0.23
5.	7	6	0.32	0.54	0.37

The second table (% of Error heading spans NN and Quadratic classifier columns).

Sl. No.	NN		Quadratic classifier			
			64 feature		400 feature	
	Rejection	Error	Rejection	Error	Rejection	Error
1.	1.84	7.78	1.37	7.30	1.31	4.69
2.	6.46	4.24	5.79	4.72	6.73	1.94
3.	15.38	2.66	14.20	1.98	10.48	0.97
4.	21.01	1.96	21.23	0.97	22.20	0.07

If we compare the confusing pairs of Odia characters, the authors of [14], have done the analysis by taking curvature features fed into quadratic classifier. The overall accuracy obtained is 94.6%.

4.7 Challenges of Odia Text Recognition

It is observed that most of the literatures reported so far face the problems in Odia OCR due to the presence of large number of vowels, consonants, vowel modifiers, compound characters and numerals and in comparison the system for Odia hand-written character recognition is facing more difficulties than printed document recognition. The structure as well as the variety of shapes of scripts and the writing style also different from other script, which pose challenges and requires more customized technique to extract a better feature set for classification. Though a number of works had done in the text recognition field for other Indian languages, the similar algorithm cannot be applicable to Odia language due to inter-class dissimilarity of characters. A standard database is not present that comprises all the types like alphabet, numeral and also compound characters. An open access standard benchmark database must be created for Odia handwritten characters. Though a number of works have done still it require improvements in pre-processing step of OCR that is helpful to store the old documents written in palm leaves.

4.8 Conclusion and Future Direction

Character recognition is the machine simulation of online or offline input character's recognition. It's the ability to acquire, clean, segment and recognize the given image characters. All types of recent advances in the area of Odia handwritten character recognition are described in this chapter. Character recognition is not a single step process, it comprises of multiple sub steps and all these steps are clearly described. The quality of classification greatly depends on the feature subset and the type of classifier used. In comparison of standard evaluation metrics, k-NN results better than other classification algorithms as it takes less space and execution time. An analysis of different classifiers with their recognition accuracy for both handwritten and printed characters and the metrics used to know a good classifier is also presented. Although various techniques for identification of handwritten alphabets have been developed in previous decades, much research is still needed to make a practical software solution available and a model should be developed for complex characters (*yuktakshyara*). Different kinds of optimization algorithms should be employed to find a good feature set that indirectly dependent on a better classification result. The existing handwritten OCR has significantly less precision and in order to solve this problem, we need a competent solution so that overall performance can be increased.

References

[1] B. B. Chaudhuri, U. Pal, and M. Mitra, 2001, "Automatic recognition of printed oriya script," in Proceedings of the International Conference on Document Analysis and Recognition, ICDAR, pp. 795–799.

[2] S. Mohanty, Nov. 1998, "Pattern Recognition in Alphabets of Oriya Language using Kohonen Neural Network," Int. J. Pattern Recognition Artificial Intelligence. vol. 12, no. 07, pp. 1007–1015.

[3] U. Bhattacharya and B. B. Chaudhuri, 2005, "Databases for research on recognition of handwritten characters of Indian scripts," Proc. Int. Conf. Doc. Anal. Recognition, ICDAR, vol. 2005, pp. 789–793.

[4] K. S. Dash, N. B. Puhan, and G. Panda, 2017, "Odia character recognition: a directional review," Artif. Intell. Rev., vol. 48, no. 4, pp. 473–497.

[5] R. K. Mohapatra, T. K. Mishra, S. Panda, and B. Majhi, 2015 ,"OHCS: A database for handwritten atomic Odia Character Recognition," 5th Natl. Conf. Comput. Vision, Pattern Recognition, Image Process. Graph. NCVPRIPG 2015.

[6] K. S. Dash, N. B. Puhan, and G. Panda, 2015, "On extraction of features for handwritten Odia numeral recognition in transformed domain," ICAPR 2015–2015 8th Int. Conf. Adv. Pattern Recognition, pp. 0–5.

[7] Dash Kalyan S, Puhan N.B., 2014, "Non Redundant Stockwell Transform Based Faeture Extraction For Handwritten Digit Recognition," IEEE Int. Conf. Signal Process. Communication.

[8] T. K. Mishra, B. Majhi, and S. Panda, 2013 ,"A comparative analysis of image transformations for handwritten Odia numeral recognition," Proc. 2013 Int. Conf. Adv. Comput. Commun. Informatics, ICACCI 2013, pp. 790–793.

[9] K. S. Dash, N. B. Puhan, and G. Panda, 2016 ,"BESAC: Binary External Symmetry Axis Constellation for unconstrained handwritten character recognition," Pattern Recognit. Lett., vol. 83, pp. 413–422.

[10] S. Mishra, D. Nanda, and S. Mohanty, 2010, "Oriya Character Recognition using Neural Networks," Spec. Issue IJCCT, vol. 2, no. 4.

[11] S. Mohanty, H. N. Dasbebartta, and T. K. Behera, 2009, "An efficient bilingual Optical Character Recognition (English-Oriya) system for printed documents," Proc. 7th Int. Conf. Adv. Pattern Recognition, ICAPR 2009, no. 1, pp. 398–401.

[12] R. K. Mohapatra, B. Majhi, and S. K. Jena, 2016, "Printed Odia digit recognition using finite automaton," Smart Innov. Syst. Technol., vol. 43, pp. 643–650.

[13] M. Nayak and A. K. Nayak, 2017,"Odia character recognition using back propagation network with binary features," Int. J. Comput. Vis. Robot., vol. 7, no. 5, pp. 588–604.

[14] U. Pal, T. Wakabayashi, and F. Kimura, 2007 ,"A system for off-line oriya handwritten character recognition using curvature feature," Proc. – 10th Int. Conf. Inf. Technol. ICIT 2007, pp. 227–229.

[15] K. S. Dash, N. B. Puhan, and G. Panda, 2014, "Non-redundant stockwell transform based feature extraction for handwritten digit recognition," in 2014 International Conference on Signal Processing and Communications (SPCOM), pp. 1–4.

[16] D. Basa and S. Meher, 2011,"Handwritten Odia Character Recognition," National Conference on Recent Advances in Microwave Tubes, Devices and Communication Systems, pp. 5–8.

[17] M. Nayak and A. K. Nayak, 2015, "Odia-Conjunct Character Recogntion using Evolutionary Algorithm," Asian Journal of Applied Sciences, vol. 03, no. 04, pp. 789–790.

[18] C. Mitra and A. K. Pujari, 2013, "Directional Decomposition for Odia Character Recognition," Springer, Cham, pp. 270–278.

[19] T. K. Mishra, B. Majhi, P. K. Sa, and S. Panda, 2014, "Model based odia numeral recognition using fuzzy aggregated features," Front. Comput. Sci., vol. 8, no. 6, pp. 916–922.

[20] K. S. Dash, N. B. Puhan, and G. Panda, 2014 ,"A hybrid feature and discriminant classifier for high accuracy handwritten Odia numeral recognition," IEEE TENSYMP 2014–2014 IEEE Reg. 10 Symp., pp. 531–535.

[21] P. K. Sarangi, A. K Sahoo, and P. Ahmed, 2012 ,"Recognition of Isolated Handwritten Oriya Numerals using Hopfield Neural Network," Int. J. Comput. Appl., vol. 40, no. 8, pp. 36–42.

[22] M. K. Mahato, A. Kumari, and S. Panigrahi, 2014, "A System For Oriya Handwritten Numeral Recognition For Indian Postal Automation," IJASTRE, pp. 1–15.

[23] P. K. Sarangi, P. Ahmed, and K. K. Ravulakollu, 2014, "Naïve Bayes Classifier with LU Factorization for Recognition of Handwritten Odia Numerals," Int. J. Sci. Technol., vol. 7, pp. 35–38.

[24] U. Pal, T. Wakabayashi, N. Sharma, and F. Kimura, 2007 ,"Handwritten numeral recognition of six popular Indian scripts," Proc. Int. Conf. Doc. Anal. Recognition, ICDAR, vol. 2, pp. 749–753.

[25] P. K. Sarangi and K. K. Ravulakollu, 2002, "Feature Extraction and Dimensionality Reduction in Pattern Recognition and Their Application in Speech Recognition," J. Theor. Appl. Inf. Technol., vol. 65, pp. 770–775.

[26] D. Padhi, 2012, "A Novel Hybrid approach for Odiya Handwritten Character recognition System," IJARCSSE, vol. 2, no. 5, pp. 150–157.

[27] S. Mohanty, 2010, "A Novel Approach for Bilingual (English – Oriya) Script Identification and Recognition in a Printed Document," Int. J. Image Process., vol. 4, no. 2, pp. 175–191.

[28] M. Nayak, 2013 ,"Odia Characters Recognition by Training Tesseract OCR Engine Odia Characters Recognition by Training Tesseract OCR Engine," Int. J. Comput. Appl., vol. 1, pp. 25–30.

[29] I. Rushiraj, S. Kundu, and B. Ray, 2017, "Handwritten character recognition of Odia script," Int. Conf. Signal Process. Commun. Power Embed. Syst. SCOPES 2016 – Proc., pp. 764–767.

[30] T. K. Bhowmik, S. K. Parui, U. Bhattacharya, and B. Shaw, 2007 ,"An HMM based recognition scheme for handwritten Oriya numerals," Proc. – 9th Int. Conf. Inf. Technol. ICIT 2006, pp. 105–110.

[31] P. K. Sarangi and P.Ahemad, 2013, "Recognition of Handwritten Odia Numerals Using Artificial Intelligence Techniques," Int. J. Comput. Sci. Appl., vol. 2, no. 02, pp. 41–48.

[32] B. Majhi, J. Satpathy, and M. Rout, 2011 ,"Efficient recognition of Odiya numerals using low complexity neural classifier," Proc. – 2011 Int. Conf. Energy, Autom. Signal, ICEAS – 2011, pp. 140–143.

[33] N. Tripathy, M. Panda, and U. Pal, 2004, "System for Oriya handwritten numeral recognition," SPIE Proc., vol. 5296, pp. 174–181.

[34] K. Roy, T. Pal, U. Pal, and F. Kimura, 2005, "Oriya handwritten numeral recognition system," in Eighth International Conference on Document Analysis and Recognition (ICDAR'05), Vol. 2,pp. 770–774.

[35] T. K. Mishra, B. Majhi, and R. Dash, 2016 ,"Shape descriptors-based generalised scheme for handwritten character recognition," Int. J. Comput. Vis. Robot., vol. 6, no. 1/2, pp. 168–175.

Arati Deshpande and Emmanuel M.

5 Pre Filtering with Rule Mining for Context Based Recommendation System

Abstract: Recommendation systems have been the integral part of web and mobile applications in the domains of e commerce, e learning, e health, e governance, social networking and search engines. The problem of spending more time in getting the required relevant information from the many options available, also called as the information overload problem is addressed by the recommendation systems. The context based recommendation systems are the types of recommendation systems which use the context information to provide the recommended items. The context is the data about the application or the surroundings or the purpose with which the user is interacting with the system which can be like time, location, type of product, user's purpose or any situation describing the interaction. In this work the architecture of the context based recommendation system is proposed with the pre filtering method with context rules. The class and object based model of context with rules and recommendation system is proposed which can be converted into a relational model for data storage for the recommendation system. One of the actions like rating is used to predict the preference of items for the current user. The analysis of the system is carried out with the Het Rec 2011 Movielens data set. The accuracy measure MAE (Mean Absolute Error) is analysed in the proposed work and the relevance is measured in terms of precision, recall and F1 measure. The experimental result shows the influence of context on action and improvement in quality of recommendation with the proposed method.

Keywords: Context based recommendation system, Pre filtering, Context modeling, Class association rule mining, Collaborative filtering.

5.1 Introduction

The recommendation systems are the software systems with intensive data processing and prediction methods. With the development of data science and internet technologies, the recommendation systems have become part of many web and mobile applications. They suggest the items preferred by the users in future considering the user's preferences and preferences of other users, who have same interests. The items suggested can vary from e commerce products to friends in social network. The recommendation systems have been developed from 1990s with the systems for web page recommendation which have been evolved into recommendation engines in web and mobile applications [1, 2]. These are called the personalized recommendation systems as they take in to account the interests of user [3, 4, 5].

https://doi.org/10.1515/9783110610987-007

Recommendation systems address the problem of information overload and reduce the burden of the user. Research on developing recommendation systems and methods has increased with the increase in online users and information. In addition, many real-world data sets that are made available in the community (Movie-Lens, Lastfm, and Yahoo! Music) have improved the progress of research on recommendation systems. The recommended items can be products such as books, music, videos, movies, electronic goods or resources such as learning resources, papers, news, or people like friends, peers or activities such as download, watch, and connect. The examples of the websites giving recommendations for different items are books in Amazon, people in LinkedIn, music in Lastfm, movies in Netflix, friends in Facebook, tourist places in TripAdvisor, and different products in e commerce applications [6]. The providers of these websites collect interaction data of the user with the system like sessions, views, clicks, downloads, transactions, ratings, reviews, and tags applied. The users' profile also gives the interests of the user. The items recommended have different characteristics like genre of movie, author of book, price of product or artist of music. The analysis of data about user and item is used to generate a recommendation for a user. The recommendation contains the list of items which will be preferred by the user in future. This list is predicted using the analysis of data of user and item [7].

5.1.1 Elements of Recommendation System

The current recommendation systems consist of a set of elements which will generate a list of personalized recommendations to a user. Independent of the method used to generate the personalized recommendation, there are some common elements of a recommendation system [8]. These can be identified as the active user, history of information, prediction engine, recommendation list generation, and recommended list as depicted in Figure 5.1.

5.1.1.1 Active User

Each user using the application is identified by the system uniquely. The user for whom the recommendation is to be generated is called the active user. The active user referred also as current user or target user, is interacting with the system for his/her need [9]. The active user's interaction with the application and the action with the recommended items are also stored in the system for future recommendation.

5.1.1.2 History of Information

The recommendation systems use the information about user, item and/or user inter-action with the system to compute the set of items preferred by the user. The user information consists of user properties such as the user's age, address, and interests. This information of user forms the user profile. The item information consists of item properties such as item category, price, and discount. The information of item forms, the item profile. The interaction of the user with the items like giving a rating to an item, giving thumbs up to an item, downloading an item, clicking an item, tagging an item, viewing an item or searching an item also forms user profile. The data about the past interaction of the user along with user profile and item profile forms the history of information which is used for the analysis activity in recommendation system [10].

5.1.1.3 Prediction Engine

A recommendation system predicts the preference of an item for the active user. The prediction of preference of each item which is not yet seen by the active user is com-puted by the recommendation system. The explicit information of user ratings for items, implicit information about the user's interaction, content of items which are the properties of items, user's demography or combination of these are used for pre-diction of preference of items to the user [11].

5.1.1.4 Recommendation List Generation

The number of items for which the preference is computed may be more as the num-ber of items can be in hundreds or thousands. Because of this, the items are filtered as top-N recommendations from the set of items not seen by the active user. The items for which the preference is calculated are sorted according to the order of pref-erence. The top-N items which can be top-10 or top-20 are selected from the sorted list to generate the recommendation list [12].

5.1.1.5 Recommended List

Recommendation list consists of the final list of recommended items to the active user. The list may change during the interaction also, if the recommendation system considers the feedback of active user or update of the history of information. The ac-tions on the recommended list of items are also stored which will be used for recom-mendations in future [13].

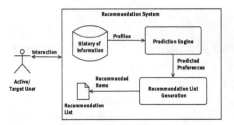

Fig. 5.1: Elements of recommendation system

5.1.2 Methods of Recommendation System

The methods used for generation of recommending items differ in the type of history of information used, prediction method used, and recommendation list generation filter. The basic methods used in recommendation systems are collaborative filtering, content filtering, and hybrid filtering. These are used in traditional recommendation methods as well as in current methods.

5.1.2.1 Collaborative Filtering

The collaborative filtering recommends to the active user, the items which are liked by other users who have similar tastes as that of the active user [14, 15]. The approach assumes that the human behaviour of persons having same interest is similar. The method to find the similar users, takes the explicit ratings of users on items in many recommendation systems. The items (which are not yet seen by the active user) which will be liked by the similar users are found by predicting the preference of the items. The similarity of users is used in user based collaborative filtering and the similarity of items is used in item based collaborative filtering.

5.1.2.2 Content Filtering

The content filtering approach recommends the items which are similar to the past preferences of active user. The items which are similar to the items liked by the active user in the past are recommended. The similar items are computed by taking the characteristics of items which form the item profile or the content of item [16, 17]. As only similar items to the active user are recommended, it may not give diverse recommendations.

5.1.2.3 Hybrid Filtering

The hybrid filtering approach combines the collaborative and content approaches to find the items to be recommended [18, 19]. It utilizes the advantages of both to overcome the shortcomings of the collaborative and content filtering approaches.

5.1.3 Context Based Recommendation System

The context based recommendation systems have been developed from early 2000 with the development of context aware systems in pervasive and ubiquitous computing. The goal is to improve the quality of recommendation which increases with the recommendation of relevant items to the user. This in turn increases the trust of users in the application and benefits the service provider. The context can be user or item attributes which can be considered as user or item profile to find the similarity of user or item. Hence the context in context based recommendation is considered as the context of interaction of user with the application. It can be time, location or user behaviour which is used as additional information for generating recommendation [20, 21]. The context can be combined with the basic recommendation algorithm by any one of the approaches called as pre-filtering, post-filtering, and contextual modeling [22]. The pre filtering approach combines the context by reducing the data to be processed which is the history of information. The post filtering approach reduces the recommendation list generated by a recommendation algorithm by using the context of current user. The pre and post filtering methods can use the existing recommendation algorithm as they will use context before or after the recommendation prediction. The contextual modeling approach combines the context of the current user in the model to predict the recommendation. It modifies the existing recommendation algorithm to combine the context element for generating the recommendation.

5.2 Related Work

The pre filtering approach used in many context based recommendation system as this approach will not change the traditional recommendation algorithm applied which can be collaborative filtering, content filtering or hybrid filtering. The pre filtering method reduces the data for prediction process by filtering the data using the current context of the active user.

The recommendation with multidimensional view is used in [23], by combining context as third dimension along with user and item as first two dimensions as in the traditional recommendation system. The context dimensions are added to make the recommendation space R as the Cartesian product of n dimensions, $D_1, D_2 \ldots D_n$ given by,

$$R : D_1 \times, D_2 \times \ldots D_n. \tag{5.1}$$

The context information is modelled as OLAP hierarchies and ratings are stored in multidimensional cubes. The reduction based approach is proposed which reduces the multidimensional contextual recommendation to two dimensional user and item dimensions. The recommendation space R with user U, item I, and context dimensions C is given by,

With context $\qquad R : U \times I \times C \rightarrow Rating,$ (5.2)

Without contex $\qquad R : U \times I \rightarrow Rating.$ (5.3)

In this method the user item matrix is reduced to contain the ratings of items given in a specific context. For time as context T, $\forall(u, i, t) \in (U \times I \times T)$ the rating prediction function $\acute{R}$ is,

$$\hat{R}(u, i, t)^D_{UxIxT} = \hat{R}(u, i)^{D(Time=t)(u,i,r)}_{UxI}, \tag{5.4}$$

where $D(Time = \acute{t})(u, i, r)$ is the set of ratings for the records having time dimension value as t, which is considered as the contextual segment with time as value t. A generalized pre filter is also proposed to reduce the sparsity when the exact pre filter is used. For example, a context segment with weekday is the generalized context of a specific day like Monday. A combined approach is used where reduction based method is used only when it outperforms the user based collaborative filtering. The evaluation parameters used are MAE, precision, recall, and F1 measure with a movie recommendation dataset. The reduction based method outperforms the user based collaborative filtering for some contextual segments which can depend on the application also.

The item splitting method proposed in [24] uses context to split items with pre filtering approach. The item with different pattern of ratings in two different contexts is split into two items with the same rating as item in summer and item in winter. The item to be split is decided on the impurity criteria determined from two sample t test, two proportion z test, information gain, chi square test or random test which decides whether there is a difference in rating under the given context or not. If the ratings for an item i, are different under context $c = cj$ and $c \neq cj$ then the item is split into two with one having ratings of item i when $c = cj$ and other having ratings of item i, when $c \neq cj$. The rating predictions for all items not rated by the target user are computed with modified *User* X *Item* matrix. This is evaluated with dataset generated for movie and yahoo dataset with age and gender for context, for MAE, precision, and recall. It is found to be beneficial and depends on the method to determine which item to split and relevance of context. The context based approach with reduction and item splitting have better accuracy than context free approach.

The 'time' is the commonly used context dimension in many context based recommendations. A pre filtering approach with context as time is used in [25]. Three 'x' months duration from current date (time) is used to define the contextual segments for prediction. The rating is predicted by collaborative filtering in three contextual segments to obtain the three lists of recommendations. The lists are combined with the weights given by Fuzzy Inference System (FIS) to obtain the recommendation for popularity. The context is given by the two input variables item popularity and user participation and recommendation is the output variable for FIS. The output weight value given by recommendation for each context is used to calculate the average prediction of rating for an item. Here $r_{u,i}$ is average prediction of rating of user u for item i is given by,

$$r_{u,i} = W_{c_1,i} r_{c_1,u,i} + W_{c_2,i} r_{c_2,u,i} + W_{c_3,i} r_{c_3,u,i} W_{c_1,i} + W_{c_2,i} + W_{c_3, i}, \tag{5.5}$$

where $r_{c_j,u,i}$ is the prediction of rating for the item i for user u in context c_j and $w_{c_j,i}$ is the weight for recommendation obtained by FIS for the item i in context c_j. The evaluation on Movielens dataset increases accuracy, as only relevant data is taken for the recommendation.

A pre filtering approach with clustering is used in [26] for scalable context aware recommendation system. The users are clustered with hierarchies according to their demographic values as context, before collaborative filtering to reduce the size of user item matrix. The users are split hierarchically, with level one cluster as Male and Female and level 2 clusters within level 1 are on age or occupation. The top-K similar users belonging to the active user's cluster are taken for recommendation for collaborative filtering. The top-N items are recommended from the item list of top-K users. The Movielens dataset is used for evaluation with MAE, precision, and recall. The recall is better compared with the context free method. The collaborative filtering run time performance increases by a factor of k if k equal partitions of users are created but the quality may reduce as the ratings of only one cluster are used in prediction.

The comparison of pre and post filtering approaches is conducted in [27]. The exact pre filtering approach is compared with weight and filter approaches of post filtering. In exact pre filtering the users' ratings in the exact context of the target user are taken for collaborative filtering. The e-commerce dataset with time of year and Amazon transactions with intent of purchase were used for evaluation. The post filtering with filter method has better performance than pre filtering. Post filtering with good filter can give better results.

There are number of measures which are used to evaluate the performance of various recommendation algorithms. The quality of recommendation depends on prediction accuracy, relevance and efficiency of the system. Statistical accuracy metrics measure the difference between the predicted rating and actual rating [28]. Mean absolute error (MAE) measures the deviation of the predictions generated by the recommender system to actual values. The MAE for each user i is calculated for n items and average of all MAE for m users is taken. Lower MAE corresponds to accurate recommendation. Given $ar_{i,j}$ as predicted rating and $r_{i,j}$ as actual rating,

$$MAE = \sum_{i=1}^{m} \left(\sum_{j=1}^{n} |ar_{i,j} - r_{i,j}| \big/_{n} \right) m. \tag{5.6}$$

Root mean square error (RMSE) is the square root of the average of square of loss of absolute error over the whole test set.

For top-N items, the need is to know whether the user will prefer some or all items in the list, to evaluate the value of the list. This can be measured with precision, recall, and F1. The dataset is divided into two training and test disjoint sets. The recommendation algorithm is applied on the training set to generate the top-N set. The items in the test set and items in the top-N, which are same from the hit set or relevant items. Precision is the ratio of the number of items relevant in the top-N set to the number of top-N recommendation. Recall is the number of relevant items in top-N set to the total number of test set items. F1 is the harmonic mean of precision and recall which is given by,

$$F1 = 2*precision*recallprecion + recall. \tag{5.7}$$

5.3 Context Modeling

A context is used in pervasive and ubiquitous computing to process the information and respond according to the situation or environment of the system. Context is also used in a software system to adapt the software functionality according to the context. The example of context can be the time of rating or location of the user or intent of purchase. The attributes of user like age or interest and attributes of items like price or discount can also be taken as context in some systems. Many researchers define context as applicable to specific systems, which make use of information like location, temperature, and user. A widely accepted definition was provided in [29] as

> "Context is any information that can be used to characterize the situation of an entity. An entity is a person, place, or object that is considered relevant to the interaction between the user and an application, including the user and applications themselves."

The context types are categorized as Activity, Identity, Location, and Time. The two views of context, namely a representational view and an interactional view is proposed in [30]. In interactional view the context is considered as the part of the activity. The context is related to activity and activity also decides the context. For example, when a user searches for a book, the context of the user's age and intent of purchase influences the search activity for type of book and also searching a particular category of books can decide the context like age or purchasing for children. The context and its values are defined by the activity, dynamically instead of statically defining the context before the processing of information or activity.

In representational view the context is a form of information, which can be extracted before processing, which is stable, and it is separated from the activity which is happening. The contexts are identified before the system design. The design and implementation of the system are done by considering the identified contexts. For example, location of user, age of a person or time can be the contexts identified in the tourist place recommendation. A system is context aware if it uses context to provide relevant information and/or services to the user, where relevancy depends on the user's task [31]. For the design of intelligent systems, the designers will have to decide the relevant context before implementation which is an important task. This representational view of context is used in this recommendation system.

The recommendation process may need the context in a different form than it is acquired from the user interaction. The information of time stamp is to be converted into day, date, and time or season for the recommendation process. A hierarchical form of representation of context is given in [32]. The context modeling methods from pervasive and ubiquitous computing are used in recommendation systems. The proposed model uses the object oriented model, to model the context and extends it with the recommendation system and user interaction. The object model of context can be converted into a relational database model to store and access the context for the recommendation process.

The context used for the recommendation system is identified at the design time and is stored as an attribute along with rating or any user interaction attribute. The context like time, location or intent of purchase is considered as the dimension added to user and item dimensions in a recommendation system [22]. A context dimension consists of contextual elements which are related in a hierarchical way and can take values from a specific domain. A context dimension hierarchy can have different hierarchical structure as depicted in Figure 5.2. The selection of variable from the hierarchy for each context dimension and their combination depends on the application.

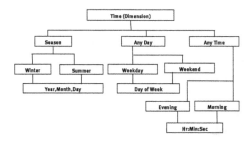

Fig. 5.2: Time context hierarchy

The context is modelled as a class with types as user, item, system, and other context in [33]. In this each context type has variables for each dimension and value pair. This model uses the object oriented concept for modeling the rating of one user on an item which is associated with zero or more of the four context types defined. This is a more generic model for recommendation using a context. This model has not considered all users and hierarchy of context dimension. The proposed model incorporates the system of recommendation with the actions and hierarchy of context dimension. It is converted into conceptual database model for storage and access of context and recommendation.

5.3.1 Conceptual Model of Context Based Recommendation System

A conceptual model of context based recommendation is proposed in terms of an object oriented model for recommendation with context and user action [34]. The object oriented model properties which are persistent can be converted into a relational database model for storage and access. The problem of top N recommendation is defined as estimating the ranking or rating of an item by a user u which he/she has not seen and extracting the top N items.

The recommendation system is defined as $RS = \{U, I, A, C, RU\}$ where U = set of users, I = set of items, A = set of actions, C = set of context dimensions, and RU = set of rules.

$U = \{u_1, u_2 \ldots u_m\}$, where each u_k defines a user profile with a set of attributes which can be $u_k = \{id, name, dob\}$, for attributes id, name, and date of birth respectively, where $1 <= k <= m$ and m = number of users.

Similarly $I = \{i_1, i_2 \ldots i_n\}$, where each i_k defines an item profile with a set of attributes which can be $i_k = \{iid, iname, icat, iprice\}$, for attributes item id, item name, item category, and item price respectively, where $1<= k <= n$ and n = number of items.

$C = \{c_1, c_2 \ldots c_d\}$, where each c_k is $c_k = \{id, name, elements\}$, for attributes context dimension id, context dimension name, and context dimension elements respectively. The elements of a context dimension c_k can have a hierarchical structure with $elements = \{c_{de1}, c_{de2}, \ldots c_{dev}\}$, where each c_{dev} defines the element v for context dimension d. Each c_{dev} takes a value from a predefined set of domain values.

$RU = \{ru_1, ru_2, \ldots ru_l\}$, where each ru_k defines a rule such that $ru_k = \{c_{de1}, c_{de2}, \ldots c_{dex}, a\}$ where c_{dex} is a context element ex for a context dimension d and a is the action type which can be characterized by some value and $1 <= k <= l$ where l = number of rules.

The recommendation problem is defined as predicting the utility UF of an item i for a user $u \in U$ or each of the item $i \in I$ in current context C' where C' is the context for current user and i is not seen by user u. This prediction is the function P of $U, I, C, A,$ and RU which are stored for the recommendation system RS and is given by,

$$UF(u, i, C') = P(U, I, C, A, RU). \tag{5.8}$$

A recommendation method or algorithm *RA* uses the prediction function and gives the top N items with highest utility function values which are not yet seen by the current user.

$A = \{a_1, a_2 \ldots a_p\}$, where each a_k defines the action profile with a set of attributes which can be $a_k = \{aid, name, value, u, i, c_1, c_2 \ldots c_d\}$, for attributes action id, action name, action value, user, item, and context dimensions c_1 to c_d respectively, where d = number of context dimensions, $1 <= k <= p$ and p = number of action types.

The proposed conceptual model of the recommendation is depicted in Figure 5.3 as a UML class model. The object of class 'RecommendationSystem' generates context based recommendation with users, items, actions, contexts, and rules. The user profile is stored in an object of class 'User' and item profile is stored in an object of class 'Item'.

Each 'ActionType' is a subclass of class 'Action' which can store information of each interaction of a user with each item corresponding to that action. For example, action types can be rating action, viewing action, buying action or tagging action. Each 'ActionType' object is associated with zero or more 'ContextDimension' objects which have a set of 'ContextElement' objects, having specific values. The Rule objects define the relation of context dimension elements with type of actions. These rules can be used for generating recommendation by the recommendation method *RA*.

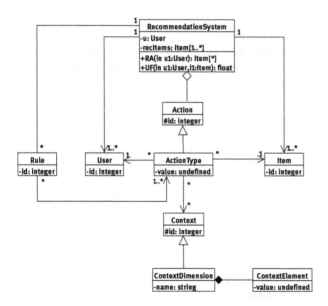

Fig. 5.3: UML class model of context based recommendation system

5.4 Architecture of Context Based Recommendation System

The context based recommendation system proposed in this work is based on the conceptual model proposed in the previous section. The general elements with the context processing elements are combined in the architecture of context based recommendation system. The overall architecture of a context aware recommender system consists of interaction manager, processing, and data repository [35]. These three elements are designed according to the applications. The presentation or interaction manager consists of user interaction with the system. The processing part consists of data access, context processing, rules processing, user or item modeling recommendation algorithm, and interaction access. The data repository consists of user, item, context, and interaction data which is accessed by the elements of the processing part of the system. The proposed context based recommendation system uses the pre filtering approach for context inclusion in the system [36]. The general architecture of the proposed context based recommendation is depicted in Figure 5.4. Each element of the system is described in next sub sections.

5.4.1 Users and Users' Actions

The users are the actors who are interacting with the system. The users are responsible for input actions such as request for recommendation or feedback actions on the items and recommendation list like rating or tagging.

5.4.2 Web Server

The web server is the application server which handles the requests and responses from web client. This server hosts the web application with the context based recommendation system. The context based recommendation system is designed as a web application with three tier client server architecture. The recommendation application hosted interacts with the database server which stores the data.

5.4.3 Context Rules Generation

The data repository stores the context information after pre processing as required for the recommendation. It also stores the user profile, item profile, and the user interaction data like rating. The context rules are generated offline which are stored in the data repository. The rules are generated using the class association rule mining. The context rules are used to select the users' action data like rating on items to apply collaborative filtering. If rating is taken as action, then the user item rating matrix is reduced before collaborative filtering using the rules generated according to context. The rules are generated offline. This rule generation is applied for any type of action to be considered for recommendation like rating or any interaction which is identified as the part of design of recommendation system.

5.4.4 Pre Filter with Context Rules

The pre filtering approach filters the user, item, and action data to be processed for recommendation according to the current context of the target user. The Pre filter with Context Rules takes the current context of active user and matches with the rules in rules database. If matching is found, the rule is applied to filter user item and action data. If no match is found, the whole data without filtering is used for recommendation which reduces the method to recommendation without context. The filtered data is used to recommend the items using collaborative filtering. This is extension to reduction based approach using pre filtering rules [22, 23].

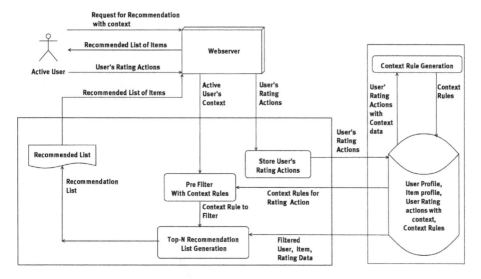

Fig. 5.4: Architecture of context based recommendation system

5.4.5 Top-N Recommendation List Generation

The Top-N Recommendation List Generation generates the top-N items to be recommended to active user using any existing collaborative filtering algorithm. The user based or item based recommendation is used for rating action in the design and implementation of the system.

5.4.6 Pre Filtering Approach with Context Rules

The architecture of the context based recommendation system includes the different algorithms for the design of the system elements. There are two main parts. The offline part generates the context rules from the data and stores in the database. The rules are generated as association rules of context elements and action [37]. The online part generates the recommendation using context rules, collaborative filtering, and recommendation list generation. The offline computation of rule generation is part of building the model for recommendation. The rules generated are stored with user, item, and context data in the data repository of the recommendation system. The model is built once and can be updated at regular intervals. The online part provides the recommendation using the model in terms of rules and collaborative filtering.

5.4.6.1 Context Rules Generation Algorithm

The context of interaction is stored with each action of the user when the user interacts with the application. The actions of user like rating which are to be processed for recommendation are identified during analysis and design of recommendation system. The context is stored as a hierarchical representation. The values of context elements of a context dimension are stored with each action. For example, context elements of the time dimension can be season and time of the day. The season can have values of summer or winter. The time of the day can have values morning or evening. The class association rules are generated with the action as the class of the rule [38]. Each rule is of the form,

$$contextvariable_1 = value_1, \ contextvariable_2 = value_2,$$

$$... \ contextvariable_k = value_k, action_i value_{i=>action_i=value_i},$$

(5.9)

where k is the number of context elements in the rule and $k <= n$, if n is the total number of context variables. The antecedent of the rule consists of context elements with values and the consequent has the class action with the value. The context elements are arranged from left to right in the table according to the order of preference defined at design time of the recommendation system. For example, if Time and Age are considered as context dimension, Time: Season, Time: Weekday, Time: DayTime, Age: Adult can be the context elements. Each context element has values defined for that attribute. For example, Time: Season = summer and Age: Adult = young are values of attributes Season for context dimension Time and Adult for context dimension Age respectively. In movie dataset, time dimension can have Time: Season, Time: Weekday, and Time: Day Time.

The class action in the consequent of a rule can be like rating, downloading or tagging. The action is stored as the last attribute in the table and the values of action will define the class of the rule. For example, for rating action, the class is rating, and class values are rating = positive and rating= negative. If the rating values are numeric, the values are converted into categorical values [39]. If the rating values are in the range from 0.5 to 5, rating is considered as positive, if the value is greater than or equal to 2.0 and rating is considered as negative, if the value is less than 2.0.

The context rules are generated and stored in rule table which are used for pre filtering. The attributes of rule table are context elements and action and each tuple in rule table is attribute values of context elements and action. The contextual segment is the users, items, and action (rating) values in a specific context. The antecedent of the rule gives the specific context to select the action values for a contextual segment. The context segment is created for each of the rule stored in rule table.

If the number of action values (rating) in a contextual segment for a given rule is less than the threshold, the rule is discarded. The rules are stored in descending order of the number of action values (rating) in their corresponding contextual segment. The rule generation uses the class association rule mining which generates the rules by calculating the rule itemsets and pruning rules [40]. The context rules generation algorithm is presented below.

Algorithm: Context Rules Generation
Input: Context and action data, minimum support and confidence, threshold for number of action values.
Output: Context Rules stored in Rules table, Contextual segment for each rule
1. Arrange the important context dimension attributes from left to right.
2. Arrange the action as the right most element in the table.
3. Generate the Class Association rules with minimum support and confidence [38].
4. Arrange the rules with given support, confidence in descending order.
5. Select the rules as k attribute predicates as antecedent and action=value as consequent.
6. If any rule with its segment (User X Item matrix) does not have enough number of action values (rating values) then discard that rule.
7. Sort the rules according to number of action values in (rating values).
8. Select the top-K rules.
9. Store the rules in Rule table.
10. Create contextual segment (*User* X *Item* matrix) for each rule and use these segments for rating prediction or preference prediction in online recommendation.
11. End.

5.4.6.2 Proposed Pre Filtering with Context Rules and Recommendation Generation Algorithm

The recommendation is generated using the pre filtering approach with extension to reduction based algorithm [23]. The rating matrix or the action data is reduced according to the context rule which matches the current context of the active user. The current context is extracted with the request for recommendation from the active user. The context of active user is converted into context element and values which are used in rule generation input table.

A matching rule for the current context is extracted from the rule table using context rule matching algorithm. The contextual segment matching that rule is taken as the input to collaborative filtering algorithm for top-N recommendation generation. The recommendation lists are generated for each action separately with the proposed pre filtering approach and context rules. The rule matching algorithm is presented as follows.

Algorithm: Rule matching
Input: Rules stored in Rules table, Current context of active user as tuple, *Contextele1 = value1, Contextele2 = value2… Contextelek = valuek*, where 1…k are the context elements in the same order of context rule generation input table.
Output: Matching rule with its Contextual segment
1. For each rule in Rule table do
    a. For 1 to k antecedents of a rule, do
        I.   Check the active user context element value for this antecedent predicate.
        Ii.   If values are equal or the context element value in rule is null (xx) go to step 1.a.
        Iii. If not equal go to next rule in step 1.
    b. If all antecedents are equal with active user context choose the rule and go to step 3.
    c. Else go to next rule from step 1.
2. If no rule is matched discard the context and use the whole user item matrix for collaborative filtering and go to step 4.
3. Else
    a. For matching rule select the created context segment and used for collaborative filtering.
4. End.

5.5 Experimental results

The configuration of the system used for implementation is Windows 7 operating system with 8 GB RAM with i5 processor. WEKA 3.8 [41] is used for the class association rule mining and Librec 2.0 [42] is used for evaluation of parameters for the recommendation system. The data set used is hetrec2011-movielens-2k. It is an extension of MovieLens10M dataset, published by GroupLens research group [43]. The dataset consists of 2113 users and 10197 movies. Each user has user id. Each movie has movie id, title, year, genre, director, actor, country, and location. Users have rating and tag assignments for the movies. The tag assignments tuples <user, tag, movie> are 47957. The rating assignments are 855598 with tuple <user, movie, rating>. Each rating and tag assignment has date and time of rating and tagging. The items to be recommended are movies. The dataset is sampled with number of users as 150 and number of movies as 200. The movies having tags less than or equal to 100 and having ratings from 20 to 200 are selected These 150 users and 200 movies with their 5109 ratings are used as the sampled dataset for the experiments.

The context rules for the dataset for rating action are generated. The generalized context elements of 'Time' dimension are considered as season, weekday (day is weekend or weekday), and time. The values of context elements are,

$$season = \{ summer, \quad winter, \quad autumn, \quad spring \},$$
$$weekday = \{ weekend, \quad weekday \},$$
$$\text{and } time = \{ morning, \quad noon, \quad evening \}.$$

The generalized context values are derived from the specific values of date and hour: minutes: second data of the dataset with rating action. It reduces the sparsity and increases the accuracy. The context rules are generated for the rating action. The attributes of table for the generation of rules are selected as season, weekday, time, rating for rating action as shown in Table 5.1. The rating attribute is numerical having scale of 0.5 to 5. It is discretized into positive and negative values with *rating = {positive, negative}*. The dataset is divided into train set and test set in the ratio of 80:20. The rule matching algorithm is used to match the rule for recommendation generation of each user in test set. The train set is reduced according to the rule matched and used for recommendation of user in test set. The reduced dataset of train set is used for collaborative filtering recommendation. The collaborative filtering with user based KNN and item based KNN is used with context. The similarity methods used are Pearson correlation and cosine measures. The sample rating matrix is shown in Table 5.2 for userKNN and itemKNN with context.

The recommendation is evaluated for accuracy and relevance. The accuracy measure used is MAE. The relevance measures used are precision, recall, and F1. The nearest neighbours are selected for userKNN and itemKNN varying from 10 to 50 in intervals of 10. The top-N value is selected as 5 to 20 in intervals of 5 for precision, recall, and F1.

Tab. 5.1: Data table for rating rule generation

Attributes	season	weekday	time	rating
Values	Winter	weekend	evening	positive

The rule generation is applied for the train set and rules are generated for minimum support of 0.2 and minimum confidence of 0.9.The number of rules generated is 9 with minimum threshold of 800 ratings in each contextual segment for rating. The rules are stored in rule table as shown in Table 5.3.

The collaborative filtering algorithms used are userKNN and itemKNN. The evaluation is done with both PCC and cosine measure for similarities. The experimental results for MAE are shown in Table 5.4. The MAE is compared with all the methods of recommendation without context given in the previous section. The graph is shown in Figure 5.5.TheMAE for the proposed itemKNN with cosine and with context is less (more accuracy) compared to all the, with context methods. It is near to UserKNN cosine without context. The MAE is less (more accuracy) for itemKNN with cosine for the without context method when compared with all the methods. The MAE for the proposed methods is slightly more when compared with, without context methods for rating action for some methods.

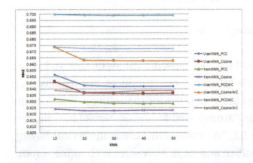

Fig. 5.5: MAE for UserKNN and ItemKNN with context for rating action

Tab. 5.2: User rating table for UserKNN/ItemKNN with context

userID	movieID	rating	seasonchar4	weekdaychar	Timechar	ratingout
75	3	1	Autumn	we	E	less
75	32	4.5	Autumn	we	E	pos
78	17	4	Autumn	we	N	pos
78	29	4.5	Spring	wd	E	pos
175	1	5	Winter	we	M	pos
175	6	0.5	Winter	we	M	Less

Tab. 5.3: Rules generated for train set of rating

ruleid	Season	Weekday	time	outclass	conf	numrates
1	xx	Wd	xx	pos	0.917	2867
2	xx	xx	E	pos	0.919	2678
3	xx	Wd	E	pos	0.919	1934

Tab. 5.3: Rules generated for train set of rating *(Continued)*

ruleid	Season	Weekday	time	outclass	conf	numrates
4	XX	We	XX	pos	0.917	1164
5	Summer	XX	XX	pos	0.928	1101
6	Winter	XX	XX	pos	0.908	1051
7	Spring	XX	XX	pos	0.911	959
8	Autumn	XX	XX	pos	0.918	920
9	XX	XX	M	pos	0.911	892

Tab. 5.4: MAE for UserKNN and ItemKNN with context for rating action

KNN	MAE			
	UserKNN_ PCCWC	UserKNN_ CosineWC	ItemKNN_PCCWC	ItemKNN_ CosineWC
10	0.700	0.674	0.674	0.639
20	0.699	0.663	0.673	0.637
30	0.699	0.663	0.672	0.638
40	0.699	0.663	0.672	0.639
50	0.699	0.663	0.672	0.639

The cosine similarity gives more accuracy for all methods. The average difference in MAE for without and with context (WC) is 0.015 and 0.043 for item KNN with cosine and PCC and 0.027, and 0.055 for user KNN with cosine and PCC respectively. The accuracy of itemKNN with cosine and context based method for rating action is near to without context based methods. The accuracy is less for this dataset as the number of ratings in each contextual segment is nearly half of the ratings used in without context methods.

Still with less than half the number of ratings, the accuracy is near to without context methods. It shows that the context based method for rating action has more accuracy with a smaller number of ratings as the ratings are contextualized.

The precision recall and F1 measures are evaluated for rating action with context at KNN=20. These are compared with the methods without context. The precision results are shown in Table 5.5 and the precision graph for the same is shown in Figure 5.6.

The precision is more for itemKNN with cosine measure with context based method compared with other methods. The userKNN with cosine for the context based method is near to itemKNN in precision. The precision for context based methods is more than the precision for without context methods. The average difference in precision for with and without context is 2% and 1.3% for item KNN with cosine and PCC and 1.7%, and 1.4% for user KNN with cosine and PCC respectively.

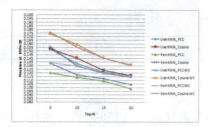

Fig. 5.6: Precision for UserKNN and ItemKNN with context for rating action

Tab. 5.5: Precision for UserKNN and ItemKNN with context for rating action

| Top-N | Precision KNN=20 | | | |
	UserKNN_ PCCWC	UserKNN_CosineWC	ItemKNN_ PCCWC	ItemKNN_CosineWC
5	0.152	0.174	0.134	0.176
10	0.133	0.159	0.129	0.155
15	0.121	0.140	0.119	0.141
20	0.116	0.131	0.115	0.130

The recall values for the context based method with rating are shown in Table 5.6 with graph in Figure 5.7.

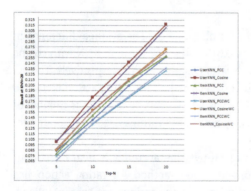

Fig. 5.7: Recall for UserKNN and ItemKNN with context for rating action

Tab. 5.6: Recall for UserKNN and ItemKNN with context for rating action

| Top-N | Recall at KNN=20 | | | |
	UserKNN_ PCCWC	UserKNN_ CosineWC	ItemKNN_ PCCWC	ItemKNN_ CosineWC
5	0.074	0.085	0.066	0.088
10	0.133	0.159	0.134	0.158
15	0.182	0.216	0.185	0.214
20	0.232	0.271	.237	0.267

The recall for userKNN with cosine in context based methods is more. The itemKNN with cosine for context based method is slightly less than userKNN with cosine and context for recall. The recall values for PCC similarity is less for user and item KNN. The recall of context based methods with rating is compared with without context methods. The recall for without context methods is more than the with context methods. The average difference in recall for without and with context is 2.2% and 1.9% for item KNN with cosine and PCC and 3.5%, and 3% for user KNN with cosine and PCC respectively. The itemKNN and userKNN cosine of context based methods have recall near to userKNN and itemKNN with PCC of without context methods. This can be because of the number of contextualized ratings in each segment.

The F1 values for the context based method with rating are shown in Table 5.7. The graph for F1 measure is shown in Figure 5.8. The userKNN with cosine and itemKNN with cosine have more relevance or F1 value for the context based methods for rating action.

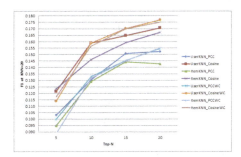

Fig. 5.8: F1 for UserKNN and ItemKNN with context for rating action

Tab. 5.7:F1 for UserKNN and ItemKNN with context for rating action

| Top-N | F1 at KNN=20 | | | |
	UserKNN_ PCCWC	UserKNN_ CosineWC	ItemKNN_ PCCWC	ItemKNN_ CosineWC
5	0.100	0.114	0.089	0.117
10	0.133	0.159	0.132	0.157
15	0.145	0.170	0.145	0.170
20	0.155	0.177	0.155	0.175

The average difference in F1 measure for with and without context is 0.9% and 0.6% for item KNN with cosine and PCC and 0.5%, and 1.5% for user KNN with cosine and PCC respectively. The PCC similarity with itemKNN and userKNN gives less relevance. The userKNN with cosine for without context is also having F1 near to the context based method as the number of ratings is more.

The proposed context based method with pre filtering for rating action has improved the relevance with itemKNN and userKNN with cosine measure for similarity. But the accuracy is near to without context method with the cosine measure for similarity. Though the accuracy is less, the number of contextualized ratings used is less than half of the non contextualized ratings.

5.6 Conclusion

Context based recommendation system with pre filtering approach with class association rule mining and rating action with collaborative filtering is proposed in this work. The major finding of this research is utilization of context information in recommendation which improves the quality of recommendation. In this work the pre filtering method uses UserKNN and ItemKNN as existing methods of recommendation. The context based recommendation system with the proposed general conceptual model can be used for the design of a context based recommendation system. The proposed architecture gives the structure of elements for the development of context based recommendation system. The context rules can be applied to extract the effect of context on user action and is applied with rating action. The contextual segments are created using context rules to apply for reduction based method for recommendation. The proposed method evaluated with the portion of Het Rec 2011 dataset and context as time increases the quality of recommendation in terms of F1 measure. Other finding is the accuracy of recommendation is not increased with the MAE measure when compared between with and without context methods, but the amount of data of user ratings, to be processed is reduced to half without much reduction in accuracy for the proposed pre filtering approach with rule mining. The relevance of recommendation in terms of precision and F1 is increased with the pre filtering approach. The approach proposed can also be used to improve the quality of recommendation as there are number of explicit and implicit actions of users which can be utilized to generate the recommendation.

References

[1] M. Pazzani and D. Billsus. Learning and Revising User Profiles: The Identification of Interesting Web Sites. Machine Learning, 27(3): 313–331, 1997.
[2] M. Pazzani and D. Billsus. Content based recommendation systems. In: The adaptive web, Springer, Berlin Heidelberg, pp. 325–341, 2007.
[3] G. Linden, B. Smith, and J. York. Amazon.com Recommendations: Item-to-Item Collaborative Filtering. IEEE Internet Computing, 7: 76–80, 2003.
[4] H. Zhang, T. Huang, Z. Lv, S. Liu, and Z. Zhou. MCRS: A course recommendation system for MOOCs. Multimedia Tools and Applications, 77(6): 7051–7069, 2018.
[5] H. C. Chao, C. F. Lai, S. Y. Chen, and H. Yueh-Min. A M learning Content Recommendation service by exploiting Mobile social interactions. IEEE Transactions on Learning Technologies, July-Sept. 7(3):221–230, 2014.
[6] J. Bobadilla, F. Ortega, A. Hernando, and A. Gutierrez. Recommender systems survey. Knowledge-Based Systems, 46: 109–132, 2013.

[7]   F. Ricci, L. Rokach and B. Shapira. Introduction to recommender systems. In: In Recommender Systems handbook, Springer, US, pp. 1–35, 2011.
[8]   M. Jallouli, S. Lajmi, and I. Amous. Designing Recommender System: Conceptual Framework and Practical Implementation. Procedia Computer Science, 112:1701–1710, 2017.
[9]   A. R. Deshpande and Emmanuel M. Recommendation System Overview. In Proceedings of 6th International conference on New Frontiers of Engineering, Science, Management and Humanities (ICNFESMH-2017), Chandigarh, ISBN: 978-93-87433-04-5, 2017.
[10] K. Verbert, H. Drachsler, N. Manouselis, M. Wolpers, R. Vuorikari, and E. Duval. Dataset driven research for improving recommender systems for learning. In Proceedings of the 1st ACM International conference on Learning Analytics and Knowledge, pp. 44–53, 2011.
[11] C. Zou, D. Zhang, J. Wan, M. M. Hassan, and J. Lloret. Using concept lattice for personalized recommendation system design. IEEE Systems Journal, 11(1):305–314, 2017.
[12] M. Deshpande and G. Karypis. Item-based top-n recommendation algorithms. ACM Transactions on Information Systems (TOIS), 22(1): 143–177, 2004.
[13] A. Elkhelifi, F. B. Kharrat, R. Faiz. Recommendation Systems Based on Online User's action. In Proceedings of the IEEE International Conference on Pervasive Intelligence and Computing, 2015.
[14] H. Jooa, S. W. Bangb, and G. D. Parka. Implementation of a recommendation system using association rules and collaborative filtering. Procedia Computer Science, 91:944–952, 2016.
[15] D. Kluver, M. D. Ekstrand, and J. A. Konstan. Rating-based collaborative filtering: algorithms and evaluation. In: Social Information Access, Lecture Notes in Computer Science, pp. 344–390, Springer, Cham, 2018.
[16] D. Wang, Y. Liang, D. Xu, X. Feng, and R. Guan. A content-based recommender system for computer science publications. Knowledge-Based Systems, 157: 1–9, 2018.
[17] P. Lops, M. De.Gemmis, and G. Semersro. Content based recommender systems. In: Recommender Systems handbook, Springer US, pp. 73–105, 2011.
[18] Z. Ali, S. Khusro, I. Ullah, and Irfan Ullah. A Hybrid Book Recommender System Based on Table of Contents (TOC) and Association Rule Mining. In Proceedings of 10th ACM International Conference on Informatics and Systems, pp. 68–74, 2016.
[19] A. Mohammed and K. Alsalama. A hybrid recommendation system based on association rules. WASET International Journal of Computer and Information Engineering, 9(1), 2013.
[20] Z. Zhang, H. Pan, G. Xu, Y. Wang, and P. Zhang. A Context-Awareness Personalized Tourist Attraction Recommendation Algorithm. Cybernetics and Information Technologies, 16(6): 146–159, 2016.
[21] K. A. Achmad, L. E. Nugroho, and A. Djunaedi. Tourism contextual information for recommender system. In Proceedings of 7th IEEE International Annual Engineering Seminar (InAES), pp. 1–6, 2017.
[22] G. Adomavicius and A.Tuzhilin. Context-aware recommender systems. In: Recommender Systems handbook. Springer US, pp. 217–253, 2011.
[23] G. Adomavicius, R. Sankaranarayanan, S. Sen, and A.Tuzhilin. Incorporating contextual information in recommender systems using a multidimensional approach. *ACM Transactions on Information Systems (TOIS)*, 23(1):103–145, 2005.
[24] L. Baltrunas and F. Ricci. Experimental evaluation of context-dependent collaborative filtering using item splitting. User Modeling and User-Adapted Interaction, 24(1–2):7–34, 2014.
[25] X. Ramirez-Garcia and M. Garcia-Valdez. A Pre-filtering Based Context-Aware Recommender System using Fuzzy Rules. Design of Intelligent Systems Based on Fuzzy Logic, Neural Networks and Nature-Inspired Optimization, pp. 497–505, Springer International Publishing, 2015.

[26] S. Datta, J. Das, P. Gupta, and S. Majumder, SCARS: A scalable context-aware recommendation system. In Proceedings of the 3rd IEEE International Conference on Computer, Communication, Control and Information Technology (C3IT), pp. 1–6, 2015.

[27] U. Panniello, A. Tuzhilin, M. Gorgoglione, C. Palmisano, and A. Pedone, Experimental comparison of pre-vs. post-filtering approaches in context-aware recommender systems. In Proceedings of the 3rd ACM conference on Recommender systems, pp. 265–268, 2009.

[28] E. Vozalis and K. G. Margaritis. Analysis of recommender systems algorithms. In Proceedings of the 6th Hellenic European Conference on Computer Mathematics & its Applications, pp. 730–745, 2003.

[29] G. D. Abowd, A. K. Dey, P. J. Brown, N. Davies, M. Smit, and P. Steggles. Towards a better understanding of context and context-awareness. In International Symposium on Handheld and ubiquitous computing, pp. 304–307, 1999.

[30] P. Dourish. What we talk about when we talk about context. Personal and ubiquitous computing, 8(1): 19–30, 2004.

[31] C. Bauer and A. K. Dey, Considering context in the design of intelligent systems: Current practices and suggestions for improvement. Journal of Systems and Software, 112:26–47, 2016.

[32] H. Mcheick. Modeling Context Aware Features for Pervasive Computing. Procedia Computer Science, 37:135–142, 2014.

[33] C. Mettouris and G. A. Papadopoulos. Contextual modeling in context-aware recommender systems: a generic approach. In Web Information Systems Engineering–WISE 2011 and 2012 Workshops, pp. 41–52, Springer, Berlin, Heidelberg, 2013.

[34] A. R. Deshpande and Emmanuel M. Conceptual Modeling of Context based Recommendation System. International Journal of Computer Applications, 180(12): 42–47, ISSN: 0975–8887, 2018.

[35] S. Inzunza and R. Juárez-Ramírez. Building Context-Aware Recommendation Systems: A Software Engineering Point of View. In 4th IEEE International Conference Software Engineering Research and Innovation (CONISOFT), pp. 175–184, 2016.

[36] A. R. Deshpande and Emmanuel M. Recommendation System with Context and User Actions, Indian Patent 201721035439 filed 6th October 2017.

[37] R. Agrawal and R. Srikant. Fast Algorithms for Mining Association Rules in Large Databases. In Proceedings of the 20th International Conference on Very Large Data Bases, pp. 478–499, 1994.

[38] B. L. W. H. Y. Ma and B. Liu. Integrating Classification and Association Rule Mining, In Proceedings of the 4th International Conference on Knowledge Discovery and Data Mining, pp.80–86, 1998.

[39] S. Tyagi and K. K. Bharadwaj. Enhanced new user recommendations based on quantitative association rule mining. Procedia Computer Science, 10:102–109, 2012.

[40] A. R. Deshpande and Emmanuel M. Pre Filtering Approach with Association Rule Mining for Context Based Recommendation. International Journal of Allied Practice, Research and Review, 5(4): 29–39, ISSN: 2350–1294, 2018.

[41] F. Eibe, M. A. Hall, and I. H. Witten. The WEKA Workbench. Online Appendix for Data Mining: Practical Machine Learning Tools and Techniques, Morgan Kaufmann, 2016.

[42] G. Guo, J. Zhang, Z. Sun, and N.Yorke-Smith. LibRec: A Java Library for Recommender Systems. In Workshop Proceedings of the 23rd Conference on User Modelling, Adaptation and Personalization (UMAP), 4, 2015.

[43] A. Cantador, P.L. Brusilovsky, and T. Kuflik, 2nd Workshop on Information Heterogeneity and Fusion in Recommender Systems (HetRec 2011). In Proceedings of the 5th ACM conference on Recommender systems, pp. 387–388, 2011.

Debarshi Mazumder, Sudarshan Nandy, and Partha Pratim Sarkar

6 Early detection of crop diseases using machine Learning based intelligent techniques: a review

Abstract: In some underdeveloped and developing countries, the identification of crop diseases depends on the farmer's field experience and may lead to a degradation in the quality of production. The production quality can be improved, if the diseases detected in an earlier stage and machine-learning techniques are applied for detection of the crop diseases. The implementation of the decision support system is also an important part because the farmer needs to know proper information in a real-time manner and hence proper action can be taken in an earlier stage. Several advanced techniques are proposed for early detection of the crop diseases and it is inspired to survey on the techniques of machine learning based decision support system, address the issues related to productivity in agriculture. Under this study, the specific model of machine learning techniques is analyzed and discussed. Our finding indicates the justification and efficiency of incorporating the machine learning techniques with a decision support system for early detection of crop diseases.

Keywords: Machine learning, Decision support system, Agriculture intelligent decision support system, precision agriculture, smart farming.

6.1 Introduction

In this era of sustainable farming, the farmers are started to practice agriculture by maintaining the environment and food quality. The overall production of the farmers should meet the human needs to bring a good life for everyone in society [1]. The extreme weather events and the effect of climatic change are harmful to productivity in agriculture. In human history, the population growth and socioeconomic changes are the reason behind the food shortage [2]. Productivity in farming is one of the main concern in sustainable farming as it balances the demand for present and future needs. Productivity becomes a more important factor and it is measured by the Total factor productivity (TFP). This TFP rate is not in good shape to double the agriculture production by 2050 and it is 1.66 per cent instead of 1.75 per cent. In the case of a lower-income country, the growth rate is declining and it is dropped from 1.5% in 2015 to 1.24% in 2017 [1]. In this alarming situation in most of the underdeveloped and developed countries, where a large population depends on agriculture, sustainable farming is one of the solutions to balance its present and future needs. The usage of the advanced mechanism in sustainable farming makes agriculture more productive and target oriented. If we consider India then Indian weather becomes hotter, drier and wetter which may cause a 10% decline in the productivity

https://doi.org/10.1515/9783110610987-008

of the major crops by 2035 [1]. This kind of changes in the weather can be seen in other countries also. In view of the above-mention situation, it is clear that farming in most of the underdeveloped countries should incorporate more advanced techniques in farming for target-oriented productivity.

Involvement of the advanced technologies in the agriculture domain is influenced and inspired by the existing challenges present in agriculture. Since 1990, many re- searchers have been introduced to various technological advancement in several stages in precision agriculture [3]. The life-cycle of site-specific crop management or preci- sion agriculture system consists of three major components and they are recognition of diseases phase, monitoring phase and the diseases controlling phase [4]. Weather or climate, pest, disease, soil properties, water supply, irrigation, etc. are the pro- duction attributes in precision agriculture. Depending on these production attributes, many computer-based advanced technologies are introduced in many agricultural ap- plications [5]. These applications are an estimation of status, identification of soil properties, estimating evapotranspiration, estimation of drought stress, fertilizer man- agement, identification of weed, detection of pests and diseases, etc. [4] [6]. Following Computer-based technologies are incorporated for those applications: Cyber-Physical System (CPS) [7], Wireless Sensor Network (WSN) [8], Global Positioning Systems (GPS) [9], Agricultural Robots (AR) [11], Internet of Things (IoT) [10, 18, 19].

A CPS based framework is designed and developed for detecting and monitoring the different types of plant stress [9]. In the agricultural environment, soil and crop management are the leading regions where WSN plays a significant role to enhance agricultural production. WSN helps to sense or collect data from agricultural environments and sending it to the cloud-based network. Examples of data or information captured by WSN are temperature, humidity, soil moisture, CO_2 and NO_2 concentra- tions, etc. [7]. Another new technology with sensors Robotic Agriculture is introduced in the different fields of agriculture to improve and enhance agricultural activities [11]. Gradually farmers are replaced by agricultural robots to perform the following oper- ations on the field such as spraying, irrigation, and selective harvesting, crop disease detection [12] [13]. Currently, Machine Learning is enhancing a high significance in the practical application and integrating with agricultural networking systems [14]. ML helps to identify the plant's diseases and pests attack [15]. Various types of machine learning techniques are applied to the different stages in the agricultural environment [16]. These stages include classification, detection and segmentation [17]. Internet of Things or IoT is represented by different physical objects or "Things". Examples of those "Things" are sensors, agricultural robots, agricultural vehicle, lights in the household and commercial environments, drone, satellite, camera, electronic appli- ances, etc. These physical objects operate on a large volume of various data with the help of machine learning methodology in distributed ways either as a group or as a single unit [18]. In the field of agriculture, IoT is organized based on three layers. The bottom-most layer is the perception layer. This layer senses data through WSN. The intermediate layer is the network layer. This layer transfers data to the cloud. Further, the topmost layer is the application layer. The function of this layer is to store, manipulate and classify the data coming from the network layer [19]. However, IoT is still booming in the domain of agriculture. Recent research in this field is showing that IoT tries to achieve its ultimate goal, which will be a communication of non-human things without any human interactions. This will lead to numerous machine-to-machine (M2M) connection in the near future [20].

IoT is leading the society towards "smart farming" [19]. Agricultural production challenges are very important for smart farming to maintain production of the crop, ecological effects, access towards nutritious food (food security) and sustainable devel- opment of a balanced environment. Smart farming relevantly and directly linked with sustainable agriculture [21] [22]. Sustainable agriculture is specified as an ecosystem- based methodology. In this methodology, three key objectives are achieved in together. Those objectives are the achievement of efficient environmental health and safety, financial cost-effectiveness and fairness in society and economy [23]. Sustainable agri- culture system is proposed in its wide-ranging sense, from the local ecosystem to each individual farm, and to the system communities affected by this farming system both locally and globally [24]. It gives the tools to explore the inter-communication between farming and other environmental aspects. Environmental aspects are bio-logical, chem- ical, physical, ecological, economic and social sciences [1] [5].

Decision support system (DSS) is a major tool in Sustainable farming to realize the farm economy with less negative environmental impact [13]. Early systems available to farmers are not using with full potential be-cause it fails to capture the actual needs of the farmers. Existing DSS are developed on those parameters, which scientists and sys- tem developers consider as essential [25]. DSS based Sustainable farming production en-compasses a variety of important factors that encourage plant productions. These factors are landscape, soil features, weather condition, and attack of pests or diseases, availability of corresponding data and aim of each farmer. In sustainable farming, Recognition and expansion of DSS assist finest governance, endorsement, set of tactics and supervision in agribusiness [21][26]. Existing DSS supported agricultural system incorporates both plants and animals. This system requires developing and adopting alternative crops or cropping systems that are suitable to the situations of farmers, weather and soil conditions of a specific area [27]. In another hand, DSS based pest or disease management are crucial to increase farm profitability and endorse the accep- tance of agricultural practices that are both sustainable and productive [23][28]. In the agricultural, the monitoring and preventing the pest attack is performed through the correct categorical identification of pest and its related diseases and this produced a satisfactory result [29]. DSS based detection and recognition system is developed for insect, pest and disease identification to prevent and control damage caused by agricul- tural pests and diseases. A classification system is also developed for insect detection. This system is incorporated a deep recursive super-resolution network using Laplacian Pyramid methodology for detecting an insect [30].

Several different types of research exist over DSS. These existing re-searches are ad- dressed different issues which solved through DSS. State of the art DSS motivates the suitable application of organic substitution in-stated of inorganic fertilizers and agrichemicals [31]. Through DSS, numerous novel approaches can be applied in agriculture. These approaches incorporate, minimum tillage, pests or dis-eases, changing crop type grow on a particular area of land, water, weeds, to achieve soil quality and soil fertility, biodiversity and wildlife [13]. In farming areas, DSS try to establish effective near field storage and construct a logistics network unambiguously. Through methodological support, DSS can enhance its capacity to manage financial risks due to climate change and minimize the pre and post illness of crop [20]. In crop production, DSS technologies recommend an ecologically friendly environment, which requires a proper utilization of land, well-timed water supply, auxiliary or supplementary plant nutrients and minimum use of pesticides.

The main objective of this chapter is to survey the recent techniques of machine learn- ing used in the agriculture field to detect crop diseases. The survey also includes a study of a decision support system which is incorporated with machine learning for early de- tection of crop diseases. The rest of the chapter is organized in four sections. In the first section, the methodology for reviewing the relevant article is presented along with two main sub-sections i.e. a detail discussion on DSS and machine-learning techniques for agriculture. The agriculture decision support system (AIDSS) and artificial neural network (ANN), SVM, K-nearest neighbor and K-means techniques are also discussed in the scope of DSS and ma-chine learning section. In the third phase, the discussion on this advanced agriculture system is presented. Finally, the chapter is concluded in the last section.

6.2 Methodology

Three steps involved in our study domain: (a) collection of related works, (b) purifying of relevant works, and (c) detailed review and analysis of this works. In the very beginning step, a search made with some keywords for journals and conference papers. These documents are published in scientific databases such as IEEE Xplore, ScienceDirect, Google Scholar, etc. Following keywords are used for searching: "Deci- sion Support System" AND ["Agriculture" OR "Smart Farming" OR "Plant Disease"] AND "Machine Learning". By using this approach, we extracted those papers that refer to DSS or machine learning in the agricultural domain. This standing survey is also scanned for interrelated work and from this struggle, 153 papers were primarily acknowledged [13] [15] [18] [19] [24]. Out of all those papers, we have considered the papers from journals or conference and books that are written in the English language only and include some machine learning techniques for the agricultural process. In the be-ginning step, 86 papers were compacted. Again, papers are scrutinized by authors depending on two factors whether they are actually based on DSS or machine learn- ing. In the final step, each paper is examined individually, identifying the problems they mentioned, considering the resolution recommended, finding the methodology and algorithms utilized, identifying the data processing techniques applied and which machine learning techniques are used to achieve a satisfactory result.

6.2.1 Decision Support System

An Agricultural decision support systems (DSS) is a well-designed software application that at first collecting the data from different near-field sensors or cameras, then processing and organizing the data and lastly com-piling or analyzing the processed data to classify feature based decision-making for agricultural management, operations, planning and security [25][26]. DSS analysis helps enterprises to detect and solve issues and problems and makes appropriate decisions. The main parts of the DSS are data acquisition and data processing, feature collection or extraction, segmen- tation, classifications and decision-making based on artificial intelligent techniques [20][23][25][27]. In the very beginning step, a huge number of raw data are collected for the system from numerous nodes or sources such as documents, knowledge from an expert, historical data etc. A DSS handles a variety of data from various sources but these data are not ready to feed directly into the system [32]. There-fore, we need some data preprocessing techniques which help to make a dataset or feeding the data accurately into the system [33]. The process of identification and detection of charac- teristics from pre-processed data and constructing a feature set is known as "feature collection or extraction". The main objective of the feature extraction process is to minimize the actual dataset by computing the fixed characteristics or properties of an object. These characteristics or properties are based on col-or, texture, histogram, geo- metric properties and shape. In the segmentation stage, feature-based data are divided or segregated into different parts. In the field of agriculture, various methodologies or approaches introduced for segmentation, based on feature information like color information, boundaries or texture of an object [29]. In DSS, most of the time color information is used for segmentation through machine learning [34]. In the classifica- tion phase, data set are divided in two part, one is used for training purpose (known as "training data set") and other is used for testing (known as "testing data set"). These data sets are fed into a classifier that correctly identified each class, generates out-puts, and determine the result [16][17][34][35]. In the field of agriculture, decreasing effi- ciency in production and cost-effectiveness are the main challenges. These challenges are raised due to climatic inconsistency and growing over the environmental effects [2]. These challenges have stimulated a journey for techniques in which systematic data can be amalgamated into the application that can help the agriculturalists in agricultural organization decisions. These applications contain Agricultural Intelligent Decision Support System (AIDSS). AIDSS helps to make agricultural knowledge more accessible to and favorable for farmers [13][20][22].

6.2.2 Agricultural Intelligent Decision Support System or AIDSS

A large number of conventional DSS exist in both industries and current researches. However, those systems are not acceptable in a satisfactory rate towards the farmers. The reasons for these draw-backs of the conventional DSSs are various. Firstly, in conventional DSS, most of the data are up-loaded or feed into the system through manual processes. These manual processes are time-consuming due to the require-ment of the preprocessing of data. Secondly, in general, conventional DSS are stand alone in nature. Nowadays, web-based connectivity through smart devices like mobile phone, computers, tablet etc. is the key requirement for any smart system. Therefore, con- ventional DSSs sometime cannot be able to cope up with these smart devices. Thirdly, most of the DSSs could not work properly when the size of data is large and those are of different types like structured data, semi-structured data and unstructured data. Finally, conventional DSSs are not incorporated into cloud-based systems. In recent days, the cloud-based system is essential for managing dis-tributed computing resources. AIDSS is a solution to the aforementioned problem. Agricultural Intelligent decision support system (AIDSS) is an IoT based system that provides AI enable cloud-based services. AIDSS sup-port precision agriculture or smart farming approaches, which incorporates inputs as weather, water, genetic, energy, landscape, human, and economic resources. AIDSS works together with these factors and provides an analysis, which can help the farmers to tackle complex prob-lems in crop production and crop security [13][19][21]. AIDSS connections the gaps between conventional decision support system and preci- sion agriculture or smart farming.

AIDSS can handle these aforementioned drawbacks of the conventional DSS in-effi- cient way. In AIDSS, Remote sensing (RS) systems are massively used to send the data directly from various sensors to a cloud-based system where AI enabled techni-ques are used for data preprocessing and analyzing in a real-time context. Thus, AIDSS can resolve the first drawback of conventional DSS. In AIDSS, the web-based smart system helps to monitor and control the catalogue of AIDSS for different stake-holders through smart devices in real time. In this way, AIDSS can handle the second drawback of the conventional system. AIDSS system is analyzed large volume of different data in a suitable way through different machine learning techniques. Thus, the third drawback of conventional DSS is man-aged in AIDSS. Final drawback achieves through cloud computing techniques. Cloud computing is a model for facil-itating suitable, on-demand system access through the network to distributed com-puting resources like networks, servers, storage, applications, and services that can be quickly provisioned and unconfined with nominal supervision effort or cloud provider communication. The Cloud computing based AIDSS shrinks the ar-rangement and handling time, improves the interaction and the collaboration be-tween the decision makers, simplifies the user- friendliness and also reduces the cost [54][55]. The overall organization of the AIDSS system is given in Figure 6.1.

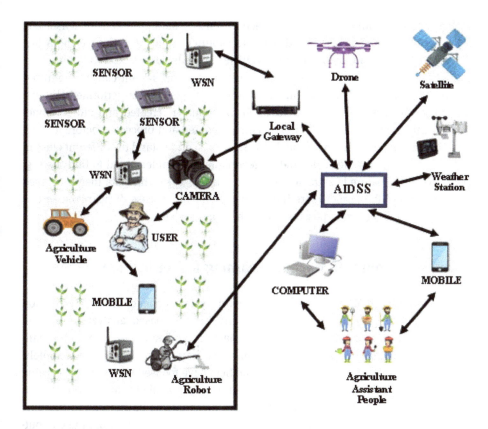

Fig. 6.1: Organization of AIDSS

AIDSS application areas are planning and preparation for food production and security, protection and finance of farmers based on weather and climate change, fertilizer management, water management and plant pests or disease identification [2] [16][15][22]. In this study, AIDSS deal with early detection of plant disease to pro- tect the crop from diseases [36]. AIDSS based Plant disease management needs data supports and that are historical climate data and statistics or symptoms about the disease, chemicals used in soil etc. [2][17]. Therefore, AIDSS can be operative if they are incorporated within bigger information systems like IoT [20]. AIDSS technique stimu- lates the periodic development of disease during the growth of the plant. So, significant action is taken while measuring the seriousness of infection [37]. This measuring depends on the availability of data. Data are either raw data or an- cient data. It is scruti- nized in the different papers that the availability of raw data instigates from numerous dissimilar sources deployed on the farmers' field [38]. In another hand, ancient data are collected by various organizations. These sources are yearbooks, non-governmental and governmental reports, regulations and guiding principle from public authorities, etc. Raw data are either image-based data or sensor-based data [39]. Data collected from sensors are bio-sensors, chemical detection devices, humidity and temperature detection sensors, sunlight intensity sensor etc. [7]. In another hand, image-based data are captured from the various camera like spectral camera, automatic airborne vehicles, mobile, drone and satellites, etc. [40][41][42]. In view of this, the proper early detec- tion system can be organized by the flowchart in Figure 6.2. In this, the data acquisition is performed through the image or sensor and the dataset in raw for-mat is created and stored some- where in the remote machine. Then data are validated and processed before feeding into the system for detection of the diseases. Now to understand the different class in between that dataset the segmentation techniques are implemented to find the seg- ment. Now the feature is extracted to classify the dis-ease. This data then processed further to make the farmer more informative about the disease of his plant. In this way, the early detection is possible through AIDSS system.

6.2.3 Techniques of Machine Learning for Agriculture

The possibilities of farming decision support system depend on data collection through remote sensing system and machine learning based analyzed report on that data [47]. In remote sensor based approaches, the key requirement is the han- dling of vast quanti- ties of sensed data generated from different platforms remotely. To manage this issue, in recent days, researchers have focused on Machine Learning (ML) methods [35]. Machine learning is used in a data analytics application and in recent years, many data analytics applications are available in the market. Machine learning provides the learning ability to a computer system that allows the comput- ers to learn the nature- based data or information automatically and make better de- cisions from experience without human interference and assistance. ML algorithms continuously improve their performance if the number of samples offered for learn- ing increases without being ex- plicitly programmed. Machine learning algorithms are categorized in four-way. These are supervised, unsupervised, semi-supervised and reinforcement. Examples of several machine learning methods are Artificial Neu- ral Networks (ANNs), Support Vector Machines (SVMs), Decision Trees, categorical and regression trees, Random Forests (RFs), K-means, k nearest neighbors, Gaussian Processes, Indian Buffet Process, etc. [12][20][31][36]. Solving enormous non-linear problems through machine learning is a great advantage when the dataset is from multiple inter-connected sources [39]. Advantages of machine learning techniques in agriculture are provided cost-effective and comprehensive solutions for the estima- tion of the quality crop, handle the problem of predicting the pre-plantation risk and decision-making for pesticide [48][51] [52] [53].

The quality and quantity of the crops are always important for the countries where a large percent-age of its population depends on agriculture. Most of the time farmers detect the pests and diseases manually due to this improper detection, the production output approximately reduced by 10% to 30% every year [1]. Good quality of crop production is achieved by timely detection of crop diseases [17].

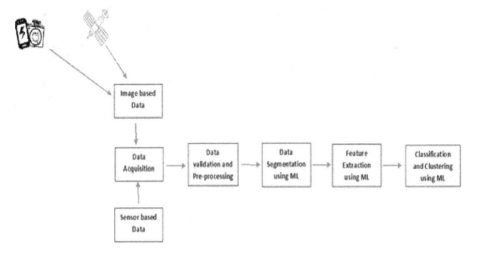

Fig. 6.2: Early detection process

The disease is mostly on leaves, roots or on the stem of the crop and it is caused by the viral, bacterial, and fungal attacks of insects, pests, rust, nematodes etc. on the plant [43][44][45][46]. It is an important task for farmers to find out these diseases as early as possible. In the recent year, many researchers are working on an AIDSS for the farmers to address the issues related to crop diseases detection and control the usage pesticide [13][29]. Usually, extensive uses of ML techniques are exhibited in the early and accurate detection of biotic stress in the crop, specifically, for detection of weeds, plant diseases and insect pests [41]. Machine learning (ML) techniques are used to represent physiological and structural features in plants. These techniques also assist to trace physiological dynam- ics due to different types of environmental effects [48][49][50]. In this article, a study is performed on limited machine learning based approach for crop diseases detection.

6.2.4 Artificial Neural Network

An artificial neuron network (ANN) is defined as a computational model that is inspired by the organization and tasks performed by a biological brain. Different types of ANN-based approaches are currently used in digital image processing techniques to detect and classify of the plant diseases through color analysis, thresholding and feature extraction [49]. Kuo-Yi Huang described a methodology that is based on back- propagation neural network classifier. In this methodology, the classifier was used to classify diseases in Phalaenopsis seedlings. These diseases are Phytophthora black rot (PBR), bacterial soft rot (BSR), and bacterial brown spot (BBS) [62]. In a recent study, the author introduced an AIDS tool based on Conventional Neural Network (CNN) to help the farmer to identify and classify the two famous banana diseases [56]. In another study, ANN architecture used to train a model on images of plant leaves. The aim of this model is classifying correctly 14 crop species and identity 26 diseases without using any feature extraction methodology [15]. Yang Lu et. al. have described a CNN based identification method for rice diseases. In this approach, CNN tries to identify 10 common diseases occurred in leaves and stems of rice plants [57]. To identify and clas- sify hyper spectral and thermal images a Back Propagation Neural Network (BPNN) based early detection ap-proach was introduced to detect biotic stresses of oil seed rape caused by fungal species [45].

6.2.5 Support Vector Machine (SVM)

Support Vector Machine (SVM) classification methods were applied in early detection of plant dis-eases for Verticillium Wilt in Olive plant. This type of classification is con- ducted using Hyper spectral and thermal Images [63]. To identify and classify maize disease a genetic algorithm based SVM technique is described in [58]. Using SVM techniques tomato plants disease are identified and classified [60, 43]. Ther- mal-based and stereo visible light-based images are also classified by SVM [60]. U. Mokhtar et. al. specified an approach that applied the SVM algorithm along with dif- ferent kernel functions to classify and identify two kinds of viruses in tomatoes leaves [43]. Caio B. Wetterich et. al. have described a custom- designed fluorescence imaging sensor for identification of HLB infection in leaf samples collected from cit- rus trees in Brazil and the USA. In this approach, normalized graph cuts are used to segment the data. Further, texture features were extracted using co-occurrence ma- trix during data preprocessing. The classifier SVM used these texture features as input [61].

6.2.6 K-Nearest Neighbors

In machine learning techniques, K-Nearest Neighbor classifier approach is very sim- ple. K-NN classifier based recognition accomplished through neighbors. These neigh- bors are nearby to query and each query class is determined using these neighbors. K-NN classifier calculates the lowest distance be-tween the given points and other points and categorizes the appropriate class. This classifier detects early-stage diseas- es existing on leaves of plants and caused by fungus. A K-NN and a BPNN together incorporated a classification method known as "Neuro-kNN" that analyze the plant leaf diseases in the primary phase before its injury [64].

6.2.7 K-means

Segmentation of plant diseases was achieved by clustering techniques. Clustering is the technique of gathering elements with similar properties based objects collected in a block or group and hence the elements inside a group is homogeneous in nature but elements from two different groups are heterogeneous in nature. Minimum Euclidean distances between the cluster center and the element are considered as a base of group- ing. In k-means clustering, k represents the total number of groups chosen previously. "Internal" and "external "are the two types of clusters exist in clustering techniques. In this effort offering a procedure that incorporates a non-supervised learning method- ology using a self-organizing map (SOM) and supervised learning based on Bayesian classifier. Images color groups are created during the training phase by SOM and group classification per-formed by K-means and finally, Bayesian classifier used for testing input image segmentation [59]. In another study, K-means algorithms used for pixel classification on digital images to identify and quantify injury on leaf surface [46].

Xia et. al. described a method that applied cloud computing to analyze visible-near infrared spectrum-based classification of apple chilling injury [65]. This specified pro- cess compared different machine learning approaches based on their performances. The outcome of this comparison revealed that the ANN classification model is more precise than the SVM classification model. However, in the case of higher-derivative spectra, the SVM method is more exact than the ANN method. Besides this, this performance comparison also exhibited that binary classification approach has lower time proficiency than a decision tree approach. However, the random forest model has higher accurateness than the decision tree model. Further, this performance evaluation also displayed that binary classification is suitable for defining an optimal breakpoint. Moreover, binary classification is a straightforward principle, which needs less estima- tion time than ANN and SVM models [65][36].

6.3 Discussion

Plant disease is the most important reason for the reduction in agricultural production. To increase the productivity of agricultural fabrication, it is of countless implication to monitoring, recognizing and controlling agricultural diseases with advanced technology. Currently, agricultural process responsibilities through computer visualization are fetching an important research area that is helpful for invisible symptoms identification on the plant. Large diversities of issues related to agricultural are presently estimated by the techniques of machine learning. Machine learning consists of a variety of altered algorithms, methods and procedures. Different ma-chine learning based disease recognition approaches were proposed in the current years. Object detection and recognition are the techniques for dis-ease recognition and it is reached by image segmentation. In reality, dis-ease images have a complex background. It is very challenging to isolate the objects from their background. Without using machine learning algorithms it is difficult to segment the complex backgrounds and has unsatisfactory achievability and operability. Most of the studies were accomplished on simple backgrounds. Therefore, the standing approaches are not very appropriate for the early detection of the disease under regular environments. This study has shown that machine learning methods for early disease detection are the only techniques through which adequate results may achieve to detect the diseases of the plant at an early stage. This review has also shown the importance of advanced techniques like the Internet of things (IoT) in the agri- culture field. The details of the AIDSS in agriculture (figure 6.1) and its justification in agriculture are described in this review. The AIDSS like systems is helpful to gather the feature from its connected sensor. Now, if the sensor is used or else the picture of the crop may be sent through the network to the remote server for the possible detection of diseases. In the remote server, it is easy for the machine learning algorithm to process and detect the diseases (figure 6.2). Hence the combination of machine learning and IoT may be very useful in case of sustainable farming and also detection of crop diseases at an early stage.

6.4 Conclusion

In the field of agriculture, loss of production due to crop diseases is one of the crucial challenges. Hence, plant disease detection has established a thoughtfulness that production quality can be improved, if the diseases are detected earlier. Different machine learning techniques are prescribed in related literature and these are very helpful in plant disease detection. Examples of these various machine learning methods are ANN, SVM, k-means clustering, K-NN etc. Individually, these machine learning method based applications are articulated for segmentation, classification and clus- ter- ing. The experimental results in disease recognition show that the projected method is a primarily appreciated approach that can support disease detection in a tiny computa- tional effort.

A postponement of this work will emphasis on the importance of incorporating decision support system along with other machine learning techniques, which may provide more accurate early detection of crop diseases. Future work also targets the area where ma- chine learning is developed only to under-stand and handle various agriculture-related natural phenomenon. Also, the decision which is deduced by the machine should be converted into human-readable text or in the local language so that the farmer can understand and cooperate with the machine.

References

[1] A. Steensland, and M. Zeigler. A World Of Productive Sustainable Agriculture. 2017 GAP report; Global Harvest Initiative; Washington, D.C.; 1–69, URL: https://www.global harvestinitiative.org/gap-report-gap-index/2017-gap-report/.

[2] P. Slavin. Climate and famines: A historical reassessment. Wiley Interdisciplinary Reviews: Climate Change, 7(3): 433–447, 2016.

[3] S. E. Cook, R. G. and V Bramley. Precision agriculture-Opportunities, benefits and pitfalls of site-specific crop management in Australia. Australian Journal of Experimental Agriculture, 38(7): 753–763, 1998.

[4] J. Bell, C. Butler, and J. Thompson. Soil-Terrain Modeling for Site-Specific Agri-cultural Management. In Proceedings of Second Inter-national Conference on Site specific management for agricultural systems, pp. 209–227, 1995.

[5] R. Bongiovanni, and J. Lowenberg-DeBoer. Precision agriculture and sustainability. Precision Agriculture, 5(4): 359–387, 2004.

[6] B. Basso, J.T.Ritchie, F.J.Pierce, R.P.Braga, and J.W.Jones. Spatial validation of crop models for precision agriculture. Agricultural Systems, 68(2): 97–112, 2001.

[7] R. Aqeel ur, A. Abbasi, N. Islam, and Z. Shaikh. A review of wireless sensors and networks' applications in agriculture. Computer Standards and Interfaces, 36(2): 263–270, 2014.

[8] W. Bastiaanssen, D. Molden, and I. Makin. Remote sensing for irrigated agriculture: examples from research and possible applications. Agricultural Water Management, 46(2): 137–155, 2000.

[9] P. Guoa, P. Dusadeeringsikul, and S. Y. Nof. Agricultural cyber physical system collaboration for greenhouse stress management. Computers and Electronics in Agri- culture, 150(2018): 439–454, 2018.

[10] R.H. Weber. Internet of Things. Springer, Springer-Verlag Berlin Heidelberg, 2010.

[11] S. Liaghat, and S.K Balasundram. A review: the role of remote sensing in pre-cision agriculture. American Journal of Agricultural and Biological Sciences, 5(1):50–55, 2010.

[12] S. J.C. Janssen, C. H. Porter, A. D.Moore, I. N. Athanasiadis, I. Foster, J. W.Jones, and J. M. Antlee. Towards a new generation of agricultural system data, models and knowledge products: Information and communication technology. Agricultural Sys- tems, 155(2017): 200–212, 2017.

[13] J. Lindblom, C. Lundström, and M. Ljung. Next Generation Decision Support Sys- tems for Farmers: Sustainable Agriculture through Sustainable IT. In 11th European IFSA Symposium, International Farming Systems Associa-tion-Europe Group, IFSA Europe. Volume 1, pp. 49–57, 2016.

[14] A. Chlingaryana, S. Sukkarieha, and B. Whelan. Machine learning approaches for crop yield prediction and nitrogen status estimation in precision agriculture: A review. Computers and Electronics in Agriculture, 151(August 2018): 61–69, 2018.

[15] S.P. Mohanty, D.P. Hughes, and M., Salathé. Using deep learning for image-based plant disease detection. Frontiers in Plant Science, 7(2016): 1–10, 2016.

[16] F.P. Luus, B.P. Salmon, F. van den Bergh, and B.T. Maharaj. Multiview Deep Learning for land-use classification. Geoscience and Re-mote Sensing Letters, 12(12):2448–2452, 2015.

[17] K. Rakesh, K.S. Amar, and R. Gajendra. Machine learning techniques in disease forecasting: a case study on rice blast prediction. BMC Bioinformatics, 7(1): 485–501, 2006.

[18] A. Tzounis, N. Katsoulas, T. Bartzanas, and C. Kittas. Internet of Things in agri- culture, recent advances and future challenges. Biosystems Engineering, 164 (2017): 31–48, 2017.

[19] A. Kamilaris, F. Gao, F.X Prenafeta-Boldú,., M.I. Ali. Agri-IoT: A Semantic Framework for Internet of Things-Enabled Smart Farming Applications. In 3rd World Forum on Internet of Things (WF-IoT), IEEE, pp. 442–447, 2016.

[20] A. Perini, and A. Susi. Developing a decision support system for integrated pro- duction in agriculture. Environmental Modelling & Software. 19(9): 821–829, 2004.

[21] M. Gafsia, B. Legagneuxb, G. Nguyenb, and P. Robina. Towards sustainable farming systems: Effectiveness and deficiency of the French procedure of sustainable agriculture. Agricultural Systems, 90 (1–3): 226–242, 2006.

[22] R., Gebbers, and V.I., Adamchuk. Precision agriculture and food security. Science, 27(5967): 828–831, 2010.

[23] B. A. Aubert, A. Schroeder, and J. Grimaudo. IT as enabler of sustainable farming: An empirical analysis of farmers' adoption decision of precision agriculture technol- ogy. Decision Support Systems, 54(1): 510–520, 2012.

[24] R., Senanayake. Sustainable Agriculture: definitions and parameters for measure- ment. Journal of Sustainable Agriculture, 1(4): 7–28, 1991.

[25] E. Jakku, and P.J. Thorburn. A conceptual framework for guiding the participa- tory development of agricultural decision support systems. Decision Support Systems, 103(9): 675–682, 2010.

[26] D.J. Power. Decision support systems: Concepts and resources for managers. Greenwood Publishing Group, Quorum Books, Westport, Connecticut, London, 2002.

[27] J. van Meensel, L. Lauwers, I. Kempen" J. Dessein. and G. van Huylenbroeck. Ef- fect of a participatory approach on the successful development of agricultural decision support systems: The case of Pigs2win. Decision Support Systems, 54(1): 164–172, 2012.

[28] D. I.Gray, W. J. Parker, and E. Kemp. Farm management research: a discussion of some of the important issues. Journal of International Farm Management, 5(1):1–24, 2009.

[29] V. Singh, and A.K. Misra. Detection of Plant Leaf Diseases Using Image Segmen- tation and Soft Computing Techniques. Information Processing in Agriculture, 4(1): 41–49, 2017.

[30] Y. Yue, X. Cheng, D. Zhang, Y. Wu, Y. Zhao, Y. Chen, G. Fan, and Y. Zhang. Deep recursive super resolution network with Laplacian Pyramid for better agricultural pest surveillance and detection. Computers and Electronics in Agriculture, 150(July 2018): 26–32, 2018.

[31] M. Pocha, J. Comas, I. Rodríguez-Roda, M. Sànchez-Marrèb, and U. Cortésb. Designing and building real environmental decision support systems. Environmental Modelling & Software, 19(9): 857–873, 2004.

[32] A. Kamilaris, A. Kartakoullis, and F. X. Prenafeta-Boldú. A review on the prac- tice of big data analysis in agriculture. Computers and Electronics in Agriculture, 143 (2017): 23–37, 2017.

[33] A. Vibhute, and S.K. Bodhe. Applications of image processing in agriculture: a survey. International Journal of Computer Applications, 52(2):34–40, 2012.

[34] L. Saxena, and L. Armstrong. A survey of image processing techniques for agricul- ture. In Proceedings of Asian Federation for Information Technology in Agriculture, Australian Society of Information and Communication Technologies in Agriculture, Perth, Australia, pp. 401–413, 2014.

[35] N. Kussul, M. Lavreniuk, S. Skakun, and A. Shelestov. Deep learning classification of land cover and crop types using remote sensing data. IEEE Geoscience and Remote Sensing Letters, 14(5): 778–782, 2017.

[36] T. Rumpf, A. K Mahlein" U. Steiner, E. C. Oerke, , H. W. Dehne, and L. Plümer. Early detection and classification of plant diseases with Support Vector Machines based on hyperspectral reflectance. Computers and Electronics in Agriculture, 74(1): 91–99, 2010.

[37] A. Mutka, and R. Bart. Image-based phenotyping of plant disease symptoms. Fron- tiers Plant Science, 5(734):1–8, 2015.

[38] J. L.Schnase, D. Q.Duffy, G. S. Tamkin, D. Nadeau, J. H. Thompson, C. M. Grieg, M. A. McInerney, and W. P.Webstera. MERRA analytic services: meeting the big data challenges of climate science through cloud-enabled climate analytics-as-a-service. Computers, Environment and Urban Systems, 61(B): 198–211, 2017.

[39] D. Waga, and K. Rabah. Environmental conditions' big data management and cloud computing analytics for sustainable agriculture. World Journal of Computer Ap- plication and Technology, 2(3): 73–81, 2014.

[40] C.H. Bock, G. H. Poole, P. E. Parker, and T. R. Gottwald. Plant diseases verity estimated visually, by digital photography and image analysis, and by hyper spectral imaging. Critical Reviews in Plant Sciences, 29(2): 59–107, 2010.

[41] J.M. Peña , J. Torres-Sánchez, A. I. de Castro, M. Kelly, and F. López-Granados. Weed mapping in early-season maize fields using object-based analysis of unmanned aerial vehicle (UAV) images. PLoS ONE, 8(10): 1–11, 2013.

[42] R. Calderón, J. A. Navas-Cortés, C. Lucena, and P. J. Zarco-Tejad. High-resolution air born hyper spectral and thermal imagery for early detection of Verticillium wilt of olive using fluore scence, temperature and narrow-band spectral indices. Remote Sensing of Environment; 139:231–245, 2013.

[43] U. Mokhtar, M. Ali, A. E. Hassanien, and H. Hefny. Identifying Two of Tomatoes Leaf Viruses Using Support Vector Machine. Information Systems Design and Intelli- gent Applications, Springer India, pp.771–782, 2015.

[44] J. Cai, Z. Zeng, J.N Connor, C.Y. Huang, V. Melino, P. Kumar, and S.J. Miklavcic. "Root Graph: a graphic optimization tool for automated image analysis of plant roots. Journal of Experimental Botany, 66(21): 6551–6562, 2015.

[45] P. Baranowski , M. Jedryczka, W. Mazurek, D. Babula-Skowronska, A. Siedliska, and J. Kaczmarek. Hyper spectral and Thermal Imaging of Oilseed Rape (Brassica napus) Response to Fungal Species of the Genus Alternaria. PLoS ONE, 10(3): 1–19, 2015.

[46] O. M. Kruse, J. Prats Montalbán, U. Indahl, K. Kvaal, A. Ferrer, and C. Futsaether. Pixel classification methods for identifying and quantifying leaf surface injury from digital images. Computers and Electronics in Agriculture, 108: 155–165, 2014.

[47] M. Ozdogan, Y. Yang, G. Allez, and C. Cervantes. Remote sensing of irrigated agriculture: Opportunities and challenges. Remote Sensing, 2(9): 2274–2304, 2010.

[48] A. Kamilaris, A. Assumpcio, A.B. Blasi, M. Torrellas, F.X Prenafeta-Boldú. Es- timating the environmental impact of agriculture by means of geospatial and big data analysis: the case of Catalonia. From Science to Society Spring-er, Luxembourg, pp. 39–48, 2017.

[49] S. Sladojevic, M. Arsenovic, A. Anderla, D. Culibrk, and D. Stefanovic. Deep neural networks based recognition of plant diseases by leaf image classification. Com- putational Intelligence and Neuroscience, 2016(6):1–11, 2016.

[50] X. Cheng, Y. Zhang, Y. Chen, Y. Wu, and Y. Yue. Pest identification via deep residual learning in complex back-ground. Computers and Electronics in Agriculture, 141: 351–356, 2017.

[51] K. Steen, P. Christiansen, H. Karstoft, and R. Jørgensen. Using Deep Learning to Challenge Safety Standard for Highly Autonomous Ma-chines in Agriculture. Journal of Imaging, 2(6):1–8, 2016.

[52] L. Deng, Y. Wang, Z. Han, R. Yu. Research on insect pest image detection and recognition based on bio-inspired methods. Biosystems Engineering, 169:139–148, 2012.

[53] J. Schmidhuber. Deep learning in neural networks: An over-view. Neural Net- works, 61: 85–117, 2015.

[54] S. Russell, V. Yoon, and G. A., Forgionne. Cloud-based decision support systems and availability context: The probability of successful decision outcomes. Information Systems and e-Business Management, Springer-Verlag, 8(3): 189–205, 2010.

[55] U. Sener, E. Gökalp, P. Erhan Eren. ClouDSS: A Decision Support System for Cloud Service Selection. Proceedings of 14th International Conference on Economics of Grids, Clouds, Systems, and Services (GECON 2017), Biarritz, France, pp.249–261, 2017.

[56] J. Amara, B. Bouaziz, A. Algergawy. A Deep Learning-Based Approach for Banana Leaf Diseases Classification. BTW workshop, Stuttgart, pp. 79–88, 2017.

[57] Y. Lu, S. Yi, N. Zeng, Y. Liu and Y. Zhang. Identification of rice dis-eases using deep convolutional neural net-works. Neurocomputing, 267(2017): 378–384, 2017.

[58] L. N. Zhang and B. Yang. Research on recognition of maize disease based on mobile internet and support vector machine technique. Advanced Materials Research, 108(13):659–662, 2014.

[59] D. Hernández Rabadán, F. Ramos, and J. Guerrero Juk. Integrating SOMs and a Bayesian Classifier for Segmenting Diseased Plants in Uncontrolled Environments. The Scientific World Journal, 2014(214674): 1–13, 2014.

[60] S. A. Raza, G. Prince, J. P. Clarkson, and N. M. Rajpoot. Automatic Detection of Diseased Tomato Plants Using Thermal and Stereo Visible Light Images. PLoS ONE, 10(4):1–20, 2015.

[61] C. Wetterich, R. Kumar, S. Sankaran, J. Belasque, R. Ehsani, and L. Gustavo Marcassa. A comparative study on application of computer vision and fluorescence imaging spectroscopy for detection of Huanglongbing citrus disease in the USA and Brazil. Journal of Spectroscopy, pp. 1–6, 2012.

[62] K. Huang. Application of artificial neural network for detecting Phalaenopsis seedling diseases using color and texture features. Computers and Electronics in Agri- culture, 57(1):3–11, 2007.

[63] R. Calderón Madrid, J. Navas Cortés, and P. Zarco-Tejada. Early Detection and Quantification of Verticillium Wilt in Olive Using Hyper spectral and Thermal Imagery over Large Areas. Remote Sensing, 7(5), 2015.

[64] J. D. Pujari, R. Yakkundimath, A. S. Byadgi. Neuro-kNN classification system for detecting fungal disease on vegetable crops using local binary patterns. Agricultural Engineering International, CIGR Journal, 16: 299–308, 2014.

[65] J. Xia, Y. Yang, H. Cao, C. Han, D. Ge and W. Zhang. Visible-near infrared spectrum-based classification of apple chilling injury on cloud computing platform. Computers and Electronics in Agriculture, 145: 27–34, 2018.

Rik Das, Sudarshan Nandy, and Siddhartha Bhattacharyya

7 Conclusion

Machine learning has been a thoroughfare in today's work-a-day world, thanks to the relentless efforts made by the scientists and researchers all across the world. Add to it the fast paced development and innovations in hardware design enabling scientists to implement the envisioned algorithms with utmost ease. This volume attempts to address the emerging trends in machine learning applications. Recent trends in information identification have identified huge scope to in applying machine learning techniques for gaining meaningful insights. Random growth of unstructured data poses new research challenges to handle this huge source of information. Efficient designing of machine learning techniques is the need of the hour. Recent literature in machine learning has emphasized on single technique of information identification. Huge scope exists in developing hybrid machine learning models with reduced computational complexity for enhanced accuracy of information identification. This volume focuses on techniques to reduce feature dimension for designing light weight techniques for real time identification and decision fusion. The salient features of the volume are aimed to enable applications of machine learning in daily lives so as to improve livelihood.

The constituent chapters address the fundamentals and advanced topics related to machine learning and its applications. The usage of supervised, unsupervised and reinforcement learning are discussed with their advantages and disadvantages. The survey of different techniques of learning process indicates that the hybridisation of learning techniques is beneficial if and only if the efficiency is improved in a modified version of the algorithm [1, 2, 3, 4, 5]. This can be surely possible if the hybridisation is performed in such a way that the disadvantage of one algorithm is treated by the advantages of another algorithm. It is always not so important to justify a classifier for its highest accuracy but it is better to look into three factors of a learning algorithm. The entire storing capacity be determined by the size of the individual module classifier itself and the size of the group (total number of classifiers exists in the group). Supervised, unsupervised and reinforcement machine learning are a description of ways in which machines or algorithms suffer on a data set. Computation cost is maximum for modern learning algorithms and memory is occupied by the data that is obviously inappropriate for several reasonable problems [6, 7, 8, 9, 10]. An orthogonal methodology is splitting the data, escaping the necessity to execute algorithms on enormous datasets. The idea of distributive machine learning includes the division into subsets of the dataset, the simultaneous learning from these sub-sets and the combination of the results. Over the moment, supervised, unchecked and strengthened teaching leads to the future of computers that are supposed to be brilliant and will help people to do everyday stuff.

Dimensionality reduction is very important for easy analysis, better visualization of information present in an image or signal or real world data. The techniques for

https://doi.org/10.1515/9783110610987-009

dimension reduction can also be combined with other techniques for getting better results. The accuracy of dimensionality reduction technique depends on the nature of the data. For easy analysis, retrieval, classification and other image processing tasks features are extracted from the images [11]. PCA [11, 12, 13], LDA [11] are some of the techniques used for dimensionality reduction of images. A. Sellami and M. Farah [14] discussed on different feature extraction methods like PCA, TLPP (Tensor Local Preserving Projection), Kernel PCA, Laplacian Eigenmaps. Dimensionality reduction techniques are also used in text mining to extract interesting and important information from unstructured text data [15]. Singular Value Decomposition (SVD) [15] is used to reduce the dimension of textual documents. TF-IDF (Term Frequency-Inverse Document Frequency) [16] is a statistical measure to find important words from a document. WET (Weight of Evidence for Text), CHI Square are some other feature selection methods [16, 17]. Expected Cross Entropy (ECE), Information Gain (IG) are also used to select interesting features [17].

The real challenge is to achieve computational intelligence in musicology is music modelling [18, 19, 20]. We humans ourselves are not able to judge the entire musical aspects in the true sense to model it for further processing. Modeling human perception and fast feature learning scalable parallel algorithms will lead to further progress in the domain. An interesting end to end applicatons for various tasks using more advanced music recognition systems is likely to dominate in the coming years. Structural approaches along with statistical methods as a combined technique may be more useful for musical applications as music is a felt phenomenon and not only just numbers. Although quantitative evaluation dominates the present research, mixed methods with combining qualitative and quantitative analysis can be more useful for music analytics. Human perception is subjective in nature and differences in perspectives lead to limited inter-rater agreement. These levels of inter-rater agreement illustrate a natural upper bound for any sort of algorithmic approach. Incorporation of human music perception and cognition in feature engineering will provide better results. Musical knowledge and pattern representation using advanced visualization techniques to model and simulate human system will advance in the coming years with more insight into human music decoding [20, 21, 22, 23]. Considering millions of available music tracks and the enormous growth of music over the Internet, tasks such as music pattern analysis and retrieval for the huge growing data will be a challenge in the coming years. Numerous innovative learning algorithms can be proposed in the near future for efficient music retrieval. Modeling human perception and fast adaptive learning algorithms will be the key to designing future intelligent systems.

Character recognition is the machine simulation of online or offline input character's recognition [24, 25, 26, 27, 28, 29]. It's the ability to acquire, clean, segment and recognize the given image characters. Character recognition is not a single step process, it comprises of multiple sub steps and all these steps are clearly described. The quality of classification greatly depends on the feature subset and the type of classifier used. In comparison of standard evaluation metrics, k-NN results better than other classification algorithms as it takes less space and execution time. Although various techniques for identification of handwritten alphabets have been developed in previous decades, much research is still needed to make a practical software solution available and a model should be developed for complex characters (yuktakshyara). Different kinds of optimization algorithms should be employed to find a good feature set that indirectly dependent on a better classification result. The existing handwritten OCR has significantly less precision and in order to solve this problem, we need a competent solution so that overall performance can be increased.

Context based recommendation system with pre filtering approach with class association rule mining and rating action with collaborative filtering is proposed in this work. The major finding of this research is utilization of context information in recommendation which improves the quality of recommendation. UserKNN and ItemKNN are notable methods of recommendation [30, 31, 32, 33].

In the field of agriculture, loss of production due to crop diseases is one of the crucial challenges. Hence, plant disease detection has established a thoughtfuness that production quality can be improved, if the diseases are detected earlier [34, 35]. Different machine learning techniques are prescribed in related literature and these are very helpful in plant disease detection. Examples of these various machine learning methods are ANN, SVM, k-means clustering, K-NN etc. Individually, these machine learning method based applications are articulated for segmentation, classification and clustering.

The volume thus comprises works related to current state of machine learning advancement and practices.

References

[1] Holte, R, 1989. "Alternative information structures in incremental learning systems", In: Machine and Human Learning – Advances in European Research, Kodrato, Y and Hutchinson, A (eds.). 121–142. Kogan Page.

[2] Fielding, A., 1999. Machine Learning Methods for Ecological Applications. Springer Science & Business Media.

[3] Bengio, Y., Learning deep architectures for AI, Found. Trends Mach. Learn. 2 (1) (2009) 1–127.

[4] Dutton, D. M., Conroy, G. V., A review of machine learning. The Knowledge Engineering Review, Vol. 12:4, 1996, 341–367

[5] Bishop, C.M., 2006. Pattern Recognition and Machine Learning. Springer, New York, USA.

[6] I.H. Witten, E. Frank, Data Mining: Practical Machine Learning Tools and Tech- niques, Morgan Kaufmann, 2005.

[7] Russell, S., Norvig, P., Artificial intelligence: A Modern Approach, Prentice Hall, Upper Saddle River, 2009.

[8] Amara, J., Bouaziz, B., Algergawy, A., 2017. A Deep Learning-Based Approach for Banana Leaf Diseases Classification. BTW workshop, Stuttgart, pp. 79–88.

[9] Deng, L., Yu, D., 2014. Deep learning: methods and applications. Found. Trends Signal Process. 7 (3–4), 197–387.

[10] Carbonell, J, Michalski, R and Mitchell, T, 1983. "An overview of machine learning", In: Michalski, R, Carbonell, J and Mitchell, T, eds., Machine Learning: An AI Approach, Morgan-Kaufmann.

[11] Shereena, V. B. and Julie, M. D. 2015. Significance of dimensionality reduction in image processing. Signal & Image Processing, An International journal (SIPIJ), 6(3), 27–42.

[12] Kavzoglu, T., Tonbul, H., Erdemir, M. Y. and Colkesen, I. 2018. Dimensionality reduction and classification of hyperspectral images using object-based image analysis. Journal of the Indian Society of Remote Sensing, 46(8), 1297–1306.

[13] Bhujle, H., Vadavadagi, B. H. and Galaveen, S. 2018. Efficient non-local means denoising for image sequences with dimensionality reduction. Multimedia Tools and Applications, 1–19.

[14] Sellami, A. and Farah, M. 2018. Comparative study of dimensionality reduction methods for remote sensing images interpretation. In 2018 4th International Conference on Advanced Technologies for Signal and Image Processing (ATSIP), 1–6. IEEE.

[15] Kumar, A. A. and Chandrasekhar, S. 2012. Text data pre-processing and dimensionality reduction techniques for document clustering. International Journal of Engineering Research & Technology (IJERT). 1.

[16] Muralidharan, A. N., Raj, N. S., Vinod, P. 2017. Dimensionality reduction techniques for text mining. In Collaborative Filtering Using Data Mining and Analysis. 49–72. IGI Global.

[17] Lu, Z., Yu, H., Fan, D. and Yuan, C. 2009. Spam Filtering Based on Improved CHI Feature Selection Method. In 2009 Chinese Conference on Pattern Recognition. CCPR 2009. 1–3. IEEE.

[18] Grosche, P., & Müller, M. Extracting Predominant Local Pulse Information from Music Recordings. IEEE Transactions on Audio, Speech, and Language Processing, 19, 1688–1701, 2011.

[19] Hu, X., Choi, K., and Downie, J.S. A framework for evaluating multimodal music mood classification. Journal of the Association for Information Science and Technology, 68(2), 2017.

[20] João Lobato Oliveira; Matthew E. P. Davies; Fabien Gouyon; Luís Paulo Reis Beat Tracking for Multiple Applications: A Multi-Agent System Architecture with State Recovery, IEEE Transactions on Audio, Speech, and Language Processing, Volume 20 Issue 10, Page 2696–2706, 2012.

[21] Lerch, A. An introduction to audio content analysis: Applications in signal processing and music informatics. Wiley-IEEE Press, 2012.

[22] Liem, C., Müller, M., Eck, D., Tzanetakis, G. and Hanjalic, A. November. The need for music information retrieval with user-centered and multimodal strategies. In Proceedings of the 1st international ACM workshop on Music information retrieval with user-centered and multimodal strategies (pp. 1–6), 2011

[23] Makarand, V. and Parag, K. Unified Algorithm for Melodic Music Similarity and Retrieval in Query by Humming. In Intelligent Computing and Information and Communication (pp. 373–381). Springer, Singapore, 2018.

[24] Chaudhuri, B. B., Pal, U., Mitra, M., 2001, "Automatic recognition of printed oriya script," in Proceedings of the International Conference on Document Analysis and Recognition, ICDAR, pp. 795–799.

[25] Mohanty, S., Nov. 1998, "Pattern Recognition in Alphabets of Oriya Language using Kohonen Neural Network," Int. J. Pattern Recognition Artificial Intelligence. vol. 12, no. 07, pp. 1007–1015.

[26] Bhattacharya, U., Chaudhuri, B. B., 2005, "Databases for research on recognition of handwritten characters of Indian scripts," Proc. Int. Conf. Doc. Anal. Recognition, ICDAR, vol. 2005, pp. 789–793.

[27] Dash, K. S., Puhan, N. B., Panda, G., 2017, "Odia character recognition: a directional review," Artif. Intell. Rev., vol. 48, no. 4, pp. 473–497.

[28] Mohapatra, R. K., Mishra, T. K., Panda, S., Majhi, B., 2015 ,"OHCS: A database for handwritten atomic Odia Character Recognition," 5th Natl. Conf. Comput. Vision, Pattern Recognition, Image Process. Graph. NCVPRIPG 2015.

[29] Dash, K. S., Puhan, N. B., Panda, G., 2015, "On extraction of features for handwritten Odia numeral recognition in transformed domain," ICAPR 2015 – 2015 8th Int. Conf. Adv. Pattern Recognition, pp. 0 – 5.

[30] Pazzani, M., Billsus, D., Learning and Revising User Profiles: The Identification of Interesting Web Sites. Machine Learning. 27(3): 313 – 331, 1997.

[31] Pazzani, M., Billsus, D., Content-based recommendation systems. In: The adaptive web. Springer, Berlin Heidelberg, pp. 325 – 341, 2007.

[32] Linden, G., Smith, B., York, J., Amazon.com Recommendations: Item-to-Item Collaborative Filtering. IEEE Internet Computing. 7:76 – 80, 2003.

[33] Zhang, H., Huang, T.,Lv, Z., Liu, S., Zhou, Z., MCRS: A course recommendation system for MOOCs. Multimedia Tools and Applications. 77(6): 7051 – 7069, 2018.

[34] Steensland, A.,,Zeigler, M., A World Of Productive Sustainable Agriculture. 2017 GAP report; Global Harvest Initiative; Washing-ton, D.C.; 1 – 69, URL: https://www.globalharvestinitiative. org/gap-report-gap-index/2017-gap-report/.

[35] Slavin, P., Climate and famines: A historical reassessment. Wiley Interdisciplinary Reviews: Climate Change, 7(3): 433 – 447, 2016.

Contributing authors

Siddhartha Bhattacharyya, CHRIST (Deemed to be University), Bangalore, India

Mamatarani Das, Utkal University, Bhubaneswar

Rik Das, Xavier Institute of Social Service, Ranchi, Jharkhand, India

Shreela Dash, iNurture Education Solutions, Bhubaneswar

Amod Deshpande, Consonance Acoustics, Aurangabad, India

Arati Deshpande, Pune Institute of Computer Technology, Pune, Maharashtra, India.

Parag Kulkarni, Iknowlation Research Labs Pvt. Ltd, Pune, India

Emmanuel M., Pune Institute of Computer Technology, Pune, Maharashtra, India.

Debarshi Mazumder, Budge Budge Institute of Technology, Kolkata, West Bengal, India.

Shashwati Mishra, Department of Computer Science, B.J.B. (A.) College, Bhubaneswar, Odisha, India.

Sudarshan Nandy, ASET, Amity University, Kolkata, West Bengal, India.

Mrutyunjaya Panda, Department of Computer Science and Applications, Utkal University, Vani Vihar, Bhubaneswar, Odisha, India.

Partha Pratim Sarkar, DETS, Kalyani University, Nadia, West Bengal, India.

Makarand Velankar, Cummins College and PICT, SPPU, Pune, Maharashtra, India.

Index

https://doi.org/10.1515/9783110610987-011

De Gruyter Frontiers in Computational Intelligence

Already published in the series

Volume 4: Intelligent Decision Support Systems
S. Bhattacharyya, S. Borra, M. Bouhlel, N. Dey (Eds.)
ISBN 978-3-11-061868-6, e-ISBN (PDF) 978-3-11-062110-5,
e-ISBN (EPUB) 978-3-11-061871-6

Volume 3: Big Data Security
I. Banerjee, S. Bhattacharyya, S. Gupta (Eds.)
ISBN 978-3-11-060588-4, e-ISBN (PDF) 978-3-11-060605-8,
e-ISBN (EPUB) 978-3-11-060596-9

Volume 2: Intelligent Multimedia Data Analysis
S. Bhattacharyya, I. Pan, A. Das, S. Gupta (Eds.)
ISBN 978-3-11-055031-3, e-ISBN (PDF) 978-3-11-055207-2,
e-ISBN (EPUB) 978-3-11-055033-7

Volume 1: Machine Learning for Big Data Analysis
S. Bhattacharyya, H. Baumik, A. Mukherjee, S. De (Eds.)
ISBN 978-3-11-055032-0, e-ISBN (PDF) 978-3-11-055143-3,
e-ISBN (EPUB) 978-3-11-055077-1